CONSTRUCTION PROJECT MANAGEMENT

A Practical Guide for
Building and Electrical Contractors

Edited by
Eddy M. Rojas

Lic., M.S., and Ph.D. in Civil Engineering

M.A. in Economics

University of Washington

Copyright ©2009 by ELECTRI International: The Foundation for Electrical Construction Inc.

ISBN 978-1-60427-002-0

Printed and bound in the U.S.A. Printed on acid-free paper
10987654321

Library of Congress Cataloging-in-Publication Data

Construction project management : a practical guide for building and electrical contractors / edited by Eddy M. Rojas.
    p. cm.
    Includes index.
    ISBN 978-1-60427-002-0 (hardcover : alk. paper)
    1. Project management. 2. Construction industry—Management.
    I. Rojas, Eddy M.
    HD69.P75.C647 2009
    690.068′4—dc22
                            2009008101

Phone: (954) 727-9333
Fax: (561) 892-0700
Web: www.jrosspub.com

To all the men and women
who build our country
with their hands, minds, and hearts.

STRATEGIC ISSUES IN CONSTRUCTION SERIES

Eddy M. Rojas, Editor-in-Chief

*Construction Productivity: A Practical Guide for Building
and Electrical Contractors*
by Eddy M. Rojas

*Construction Project Management: A Practical Guide for Building
and Electrical Contractors*
by Eddy M. Rojas

CONTENTS

PREFACE

Project management is fundamental for the success of construction companies. This book is a collection of some of the best studies commissioned by ELECTRI International: The Foundation for Electrical Construction Inc. in the area of construction project management. Studies commissioned by ELECTRI International are selected, coordinated, and monitored by some of the most progressive contractors in the construction industry and performed by outstanding scholars from some of the best universities in the United States. This combination of talents creates deliverables with valuable practical information that contractors can immediately apply and benefit from and which are based on sound methodological approaches.

This book is aimed at two distinct groups. First, it has become clear that most of the knowledge generated by the Foundation is universally applicable in the construction industry rather than exclusively to electrical construction. Therefore, in an effort to broadly disseminate its knowledge, the Foundation has decided to share some of its best studies for the benefit of the construction industry as a whole. Contractors from all trades will find value in the chapters included in this book as they provide information that can be directly applied or easily adapted to any trade. Second, the Foundation has realized that educating the next generation of leaders is vital to ensuring the success of the construction industry. Therefore, this book also aims to reach out to students in civil engineering, construction management, building technology, and similar college programs by offering practical construction concepts to complement academic lectures.

Topics include pre-construction planning, early warning signs of project distress, cumulative impact of change orders, sequencing guidelines, total quality management, quality assurance, ideal jobsite inventory levels, tool and material control systems, recommended safety practices, partnering, performance evaluations, and contract risk management.

A special feature of this book is the availability of free downloadable value-added resources in the form of practical, hands-on tools that can be applied to real-world situations. These resources are designed to enhance the learning experience as well as provide solutions to today's business challenges.

Finally, this book is part of a three-book series that disseminates ELECTRI International research studies. The other books in this series are *Construction Productivity* and *Construction Firm Management.*

ABOUT THE EDITOR

Dr. Eddy M. Rojas is a Professor in the Department of Construction Management at the University of Washington. He is also the Graduate Program Coordinator and the Executive Director of the Pacific Northwest Center for Construction Research and Education. He holds graduate degrees in civil engineering (M.S. and Ph.D.) and economics (M.A.) from the University of Colorado at Boulder and an undergraduate degree in civil engineering from the University of Costa Rica.

Throughout his academic career, Dr. Rojas has led numerous research studies in modeling, simulation, and visualization of construction engineering and management processes; engineering education; and construction economics. These studies have been sponsored by government agencies and private sector organizations such as the National Science Foundation, the U.S. Department of Education, the U.S. Army, the Construction Industry Institute, the New Horizons Foundation, and ELECTRI International. Dr. Rojas has documented and disseminated the results and findings from his research efforts in numerous publications in technical refereed journals, technical conference proceedings, and technical reports, and in invited lectures and presentations at national and international seminars, symposia, and workshops.

Dr. Rojas is well known in both academic and professional circles not only through his research and publications, but also by means of his professional activities, including his work as reviewer for the National Science Foundation, Specialty Editor for the *ASCE Journal of Construction Engineering and Management*, Chair and Technical Committee member in several congresses and conferences, reviewer for technical journals and conferences, and developer of the Virtual Community of Construction Scholars and Practitioners.

At J. Ross Publishing we are committed to providing today's professional with practical, hands-on tools that enhance the learning experience and give readers an opportunity to apply what they have learned. That is why we offer free ancillary materials available for download on this book and all participating Web Added Value™ publications. These online resources may include interactive versions of material that appears in the book or supplemental templates, worksheets, models, plans, case studies, proposals, spreadsheets, and assessment tools, among other things. Whenever you see the WAV™ symbol in any of our publications, it means bonus materials accompany the book and are available from the Web Added Value™ Download Resource Center at www.jrosspub.com.

Downloads for *Construction Project Management* include:

- Checklists for each one of the 10 categories that make up the model preconstruction planning process. In addition, standard procedures for change orders, RFIs, submittals, transmittals, billing, progress updates, a sample requirements and expectations letter and a sample letter of intent.
- A Spanish translation of Chapter 7: Recommended Safety Practices.
- A methodology for assessing contractors' TQM program progress based on the Malcolm Baldrige National Quality Award Criteria.
- Materials to assist contractors in preparing a quality assurance manual and implementing an effective quality assurance program.
- A template for a partnering workshop that includes a sample partnering agenda, an introduction to the basic partnering concepts, sample mission statements, and a sample partnering charter.
- Sample evaluation forms that contractors can customize to evaluate supervisors, journeymen, and apprentices.

PRE-CONSTRUCTION PLANNING

Dr. Awad S. Hanna, *University of Wisconsin–Madison*
Dr. Cindy L. Menches, *University of Texas at Austin*

INTRODUCTION

The importance of planning is unmistakable given the challenges faced by contractors in a competitive construction market. These challenges include reducing costs, improving labor productivity, minimizing changes, and maximizing resources to increase profitability. These challenges are intensified by increasingly tight timelines and ever more complex projects that test the management capabilities of even the best companies. A consensus exists in the construction industry that more formalized pre-construction planning is necessary to remain successful in an increasingly competitive industry. Accordingly, contractors are turning to pre-construction planning as one approach to improving their competitive edge.

The study presented in this chapter critically evaluates the relationship between pre-construction planning and project performance. As part of our research process, in-depth project-specific information was collected on the planning effort and project outcome for several successful and less-than-successful projects. This data collection effort culminated in a detailed analysis of the planning practices of these two groups to identify clear differences between successful and less-than-successful projects. Twenty-seven randomly selected companies from 11 states participated in our research effort. Data were collected on 29 successful and 27 less-than-successful projects.

The planning activities that were performed on the successful projects were used to develop a model pre-construction planning process. This model process was fashioned after those projects that performed effective planning and achieved a successful outcome. As such, this model process incorporates the best planning practices of the companies that participated in our research effort. The planning process of successful and less-than-successful projects was compared to the model process, and we discovered that those planning processes that more closely matched the model resulted in more successful performance.

METHODOLOGY

The primary goal of our research effort was to investigate and quantify the effect of pre-construction planning on project performance. To support this goal, our research attempted to provide evidence to support two main hypotheses:

1. Projects that experience an appropriate planning effort also experience more successful outcomes.
2. There are significant pre-construction planning activities that distinguish a successful project from a less-than-successful project.

The research was conducted in three distinct phases. In Phase 1, the current state of pre-construction planning was investigated by developing and administering a questionnaire to nearly 2000 randomly selected members of the National Electrical Contractors Association (NECA). Contractors were asked to identify all of the planning activities they typically performed as part of their pre-construction planning process. These data were used in Phase 2 to construct an initial model pre-construction planning process. Finally, in Phase 3 a detailed sampling plan was developed to select a random sample of willing participants to respond to a questionnaire and participate in an interview. Data were collected from those contractors who responded to the initial survey in Phase 1 and who indicated they would be willing to provide additional information about their planning practices. Data were collected on two projects per contractor—one project that was well planned and performed successfully and one that was poorly planned and performed poorly. The project characteristics, planning activities, and performance of the successful and less-than-successful projects were compared to identify distinct differences. The relationship between characteristics, planning, and performance was modeled, and evidence was provided, to support the theory that projects that experience more effective planning also tend to achieve more successful outcomes.

The data collection effort culminated in a detailed analysis of the planning practices of well-planned and poorly-planned projects to identify clear differences between successful and less-than-successful performers. The data analysis resulted in: (1) the development of a technique to classify and quantify the inherent characteristics of a project, which might influence how a project is planned; (2) the refinement and validation of a model pre-construction planning process based on the best practices of the participating contractors; (3) the creation of a scorecard to compare the actual planning effort on projects to the model planning process; (4) the development of a technique for quantifying the effectiveness of an actual planning process; and (5) the investigation and quantification of the relationship between planning practices and project outcomes.

PRE-CONSTRUCTION PLANNING PROCESS

Pre-construction planning is the planning that is performed to prepare a construction project for execution. Pre-construction planning is also referred to as execution planning, pre-job planning, and more generically as pre-planning, which is a term applied in many fields to refer to that stage of planning that occurs before an event happens. In this chapter we use the term pre-construction planning.

In general, pre-construction planning begins during the preparation of the bid and ends shortly after the project has been executed. However, the vast majority of planning activities are performed after a contractor has been notified of a pending award. Hence, this chapter covers the planning activities that are performed after notification of pending award and are generally completed before the project has been executed.

The pre-construction planning process, which resulted from a detailed analysis of the planning processes used on successful projects, consists of 46 activities classified in 10 categories. Figure 1.1 presents the main categories of the model pre-construction planning process. Team member involvement in planning will typically be driven by the type of planning that needs to be completed. Figure 1.2 presents a matrix of the 10 planning categories and potential team members who may be involved. There are several strategies for successfully implementing the model pre-construction planning process as a standard procedure on all projects:

- Top managers must reinforce their commitment to a standardized planning process and must ensure planning is performed on every project.
- The model planning process must be tailored to each project based on its particular characteristics. For example, a small project might

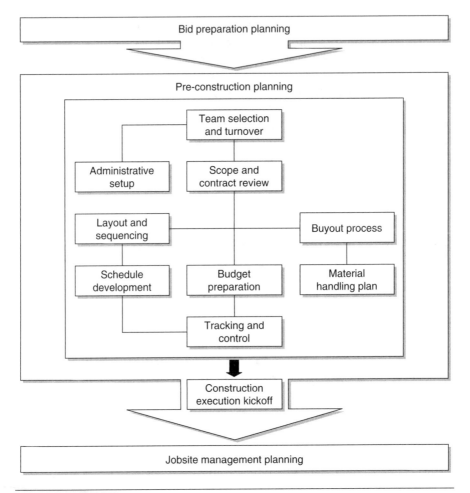

Figure 1.1 Overview of the model pre-construction planning process

require only a few minutes on some of the activities, whereas a large project might require several hours or days on those same activities.

- A system for tracking the success on projects that are well planned and that followed the model planning process will serve to document its benefits.
- In addition to implementing a model planning process, consider implementing a "double-check" system in which the supervisor double-checks the planning performed by the project manager. This should ensure planning is performed consistently across the company.

Team member

Category	Project manager	Project engineer	Estimator	Field supervisor	Foreman	CAD operator	Director of operations	Vice president	Accountant	Purchasing agent	Safety director	Admin. assistant
Team selection and turnover	✓	✓	✓	✓			✓	✓	✓	✓	✓	
Scope and contract review	✓	✓		✓			✓	✓				
Administrative setup	✓	✓							✓	✓		✓
Buyout process	✓	✓	✓							✓		
Material handling plan	✓	✓		✓								
Budget preparation	✓	✓		✓					✓			
Layout and sequencing plan	✓	✓		✓	✓	✓	✓					
Schedule development	✓	✓		✓	✓		✓					
Tracking and control	✓	✓		✓	✓				✓			
Construction execution kickoff meeting	✓	✓	✓	✓			✓	✓	✓	✓	✓	✓

Figure 1.2 Team member involvement in pre-construction planning

- The early involvement of the field supervisor in the planning process has been shown to improve the chances for a successful project. Therefore, consider selecting and involving your field supervisor during the bidding stage or in the early part of the post-award stage.
- Ensure that excellent planning effort is applied to those activities that are identified as critical. These are the activities that have the greatest potential for impacting the outcome of the project.
- The planning kickoff meeting alerts all involved team members that the planning process is officially underway, and it allows the project manager to establish deadlines for completion of activities. Likewise, the construction execution kickoff meeting brings the team members back together to review the progress of the planning and ensures that all tasks have been, or will be, completed by the deadline. These two meetings should be held on all projects regardless of size. Lessons learned and a feedback loop should be an essential part of your standardized planning process. Project managers and field supervisors should share planning practices.

The results of our study revealed that the projects that received more effective pre-construction planning were also more likely to achieve successful outcomes. Success was defined by construction contractors as follows:

1. The project was profitable
2. The customer was satisfied
3. The project resulted in repeat business
4. The project resulted in good working relationships between the trade contractor, the general contractor, and the owner
5. The worksite was safe and there were no accidents
6. The project was completed on time
7. The workers took pride in the completed project
8. There was effective communication and cooperation between the trade contractor, the general contractor, and the owner
9. The quality of the work was excellent
10, The project achieved its budget goal

Consequently, projects that are planned using the pre-construction planning process outlined in this chapter can expect to increase their chances of achieving a successful outcome. However, it must be noted that effective planning alone does not guarantee the success of a project. Instead, effective planning coupled with good project management that takes into account a project's specific characteristics improves the likelihood of achieving successful project performance. Hence, effective planning should be performed during the pre-construction stage, and good project management should be applied during the execution stage to maximize the chances of completing a project successfully.

Overall, projects that implemented a planning process similar to the model planning process presented in this chapter tended to outperform those projects that were poorly planned or whose planning process was significantly different from the model process. Figure 1.3 shows the planning effectiveness score (where a higher score indicates more effective planning) against the chances of achieving a successful outcome for the projects analyzed in this study. Projects in the upper right quadrant are those that were well planned and also performed well, while projects in the lower left quadrant were not well planned and did not perform well.

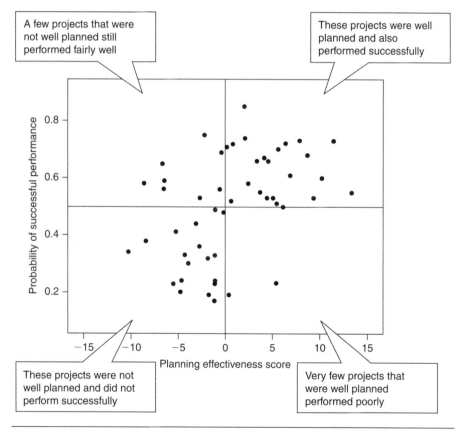

Figure 1.3 The effectiveness of planning versus the chances of achieving a successful outcome

PRE-CONSTRUCTION PLANNING ACTIVITIES

While planning begins during the bidding process, the bulk of the planning takes place during the pre-construction stage—after bidding but before execution. This section discusses those planning activities that should begin immediately upon notification of a pending award and essentially should be completed before executing the work. These activities, furthermore, set up a project to be successfully managed, which can significantly improve the chances that the project will make a profit and be completed on time.

By examining several well-planned projects, 46 activities that occurred during the pre-construction planning stage were identified and classified into 10 categories. Figure 1.4 provides a sample checklist of these 46 activities that make up the model pre-construction planning process. The activities should be checked off as they are performed to ensure that all activities are completed before the work is executed. However, there are instances when a project must begin before the pre-construction planning has been completed. The checklist can then be used to ensure that all of the planning activities are eventually completed and are not overlooked during the busy, and often intense, mobilization and execution stage.

Pre-construction planning begins immediately upon notification of a pending award. The planning may continue for several days, several weeks, or several months depending on the magnitude of the project. Planning should continue until all planning activities have been completed, even if the project must be executed prior to completing the planning process. Furthermore, the planning process should provide the framework for managing the project once the work has been executed. There are three primary goals of the pre-construction planning process:

1. To completely plan the project in a standardized and efficient manner to ensure that the tools, materials, equipment, and labor will be available to complete the project on time and within budget
2. To set up the systems that will be needed to efficiently manage the project, such as material purchasing, delivery, and storage; scheduling and tracking; change management; submittal tracking; and numerous other processes that are necessary to manage a successful project
3. To set the job up for successful execution and jobsite management

The following sections describe the 10 categories of the model pre-construction planning process shown in Figure 1.1 and the activities within each category.

Team Selection and Turnover

Team selection and turnover involves selecting the project manager and field supervisor who will be responsible for managing the job during construction execution. After the management team has been selected, a meeting should be held to formally turn over the project from the estimating stage to the project management stage. Furthermore, a planning kickoff meeting should be held to assign planning responsibilities to internal team members, including the purchasing agent, safety director, and operations director. The team selection and turnover process consists of four main activities.

Activity 1: Finalize Selection of Project Manager, Field Supervisor, and Other Key Team Members

- Assemble a list of project managers and review their current workload and special skills.
- Determine whether an ideal match can be made between the project manager and the job, and then select the project manager.
- Assemble a list of field supervisors and review their current workload and special skills.
- Determine whether an ideal match can be made between the field supervisor and the job, and then select the field supervisor.
- Work with the project manager, field supervisor, and/or the director of operations to select key foremen when appropriate.

Activity 2: Hold Turnover Meeting between Estimator and Project Manager

- Arrange a meeting time and location, and ensure that enough time has been allocated to transfer knowledge between the estimator and project manager. A formal turnover meeting is strongly recommended. The meeting time can vary widely from one hour to more than a day, depending on the size and complexity of the project.
- Follow and complete a checklist that describes the information that should be transferred. A sample turnover meeting agenda is shown in Figure 1.5.

Activity 3: Hold Separate Turnover Meetings between Project Manager and Field Supervisor

- Arrange a meeting time and location that will ensure enough time can be allocated to transfer knowledge between the project manager and field supervisor. A formal turnover meeting is strongly recommended. It is also recommended that the meeting be held at the office rather than on an existing jobsite so that the field supervisor can focus

Completed	Completion date	Category	Act. no.	Activity
☐		Team selection and turnover	1	Finalize selection of project manager, field supervisor, and other key team members
☐			2	Hold turnover meeting between estimator and project manager
☐			3	Hold separate turnover meeting between project manager and field supervisor
☐			4	Hold pre-job (planning) kickoff meeting with internal team members to assign responsibilities
☐		Scope and contract review	5	Review contract for unfavorable or high-risk clauses
☐			6	Project manager reviews plans, specifications, and schedule
☐			7	Field supervisor reviews plans, specifications, and schedule
☐			8	Create a list of issues that need to be resolved and begin the RFI process
☐			9	Conduct site visit
☐			10	Compare estimated (bid) work activities and materials to planned performance
☐			11	Identify VE and prefabrication opportunities and how to simplify the work
☐			12	Prepare construction takeoff
☐		Administrative setup	13	Set up project files and create contact list
☐			14	Set up computerized tracking and control system (forms, database, schedule, tracking)
☐			15	Initiate a change management system
☐			16	Initiate an RFI tracking and processing system
☐			17	Initiate a submittal tracking and processing system
☐			18	Develop a labor requirements/expectations letter
☐		Buyout process (continued next page)	19	Request and/or review subcontractor/supplier/vendor prices and qualifications
☐			20	Negotiate pricing and contract conditions and select subcontractors/suppliers/vendors
☐			21	Develop and issue purchase orders and contracts for materials and equipment

Figure 1.4 Checklist of pre-construction planning activities

Completed	Completion date	Category	Act. no.	Activity
☐		Buyout process (continued)	22	Order long-lead-time materials and equipment
☐			23	Request submittals, cut sheets, and shop drawings
☐			24	Develop and process log and book of submittals, cut sheets, and shop drawings
☐		Material handling plan	25	Develop material delivery and handling plan
☐			26	Develop material storage and staging plan
☐		Budget preparation	27	Develop, review, or expand cost code scheme
☐			28	Develop budget by breaking down labor, material, overhead, and profit costs
☐			29	Develop schedule of values
☐		Layout and sequencing plan	30	Develop installation sequence and layout drawings
☐			31	Develop field instructions
☐			32	Develop prefabrication drawings for field use
☐		Schedule development	33	Review general contractor's schedule and timeline
☐			34	Identify work that impacts your activities
☐			35	Review the work sequence and long-lead-time material/equipment delivery dates
☐			36	Coordinate your schedule with the schedule of the general contractor
☐			37	Create a bar chart schedule
☐		Tracking and control	38	Customize the computerized tracking and control system (e.g., database/schedule) for the current project
☐			39	Develop labor and materials tracking report
☐		Construction execution kickoff meeting	40	Review meeting schedule
☐			41	Review RFI process
☐			42	Review change order process and field change management process
☐			43	Review submittal processing procedure
☐			44	Review billing and invoicing procedures
☐			45	Review project and field reporting and tracking procedures
☐			46	Review your schedule and the schedule of the general contractor

Figure 1.4 *Continued*

1. Project overview, including:
   a. Project name
   b. Location
   c. Type of work
   d. Estimated cost and estimated work hours
   e. Profit goal
   f. General scope of work
   g. Potential opportunities and challenges of the project

2. Review the plans and specifications
   a. Go page by page through the plans to discuss the quantities and costs
   b. Review each section of the specifications
   c. Discuss how the estimator assumed the work would be performed (materials and methods)
   d. Review work performed by others
   e. Identify discrepancies and ambiguities in the bid documents
   f. Identify design discrepancies or issues
   g. Identify potential errors in the bid documents
   h. Discuss information collected during a site visit
   i. Discuss information collected during a pre-bid meeting
   j. Identify potential cost savings from changes to materials and methods
   k. Identify potential prefabrication or internal VE opportunities
   l. Discuss alternative routing of conduits
   m. Identify any temporary power and lighting requirements

3. Review the cost estimate and bid price
   a. Review each bid line item and its cost
   b. Review overhead, profit, and contingency funds
   c. Discuss wage rates
   d. Review the bid submission letter with clarifications and qualifications

4. Review subcontractor/vendor pricing and qualifications
   a. Review subcontractors and suppliers scope of work and qualifications
   b. Identify all pre-contract commitments and promises
   c. Verify minority business requirements

5. Review the schedule and milestones
   a. Review the owner-furnished schedule
   b. Review any preliminary schedule submitted with the bid
   c. Review work by others that will impact your work
   d. Identify important material and equipment delivery dates
   e. Discuss holidays, vacations, and potential weather events

6. Review manpower requirements and labor rates
   a. Review the manpower loading chart
   b. Identify the estimated crew mix
   c. Identify any wage increases
   d. Review the potential for overtime

7. Review other items specific to this project

Figure 1.5 Turnover meeting agenda

on the upcoming project without being interrupted. The meeting time can vary widely from one hour to more than a day depending on the size and complexity of the project and how many tasks must be completed during the turnover meeting.

- Follow and complete a checklist that describes the information that must be transferred. A sample turnover meeting agenda is shown in Figure 1.5.

Activity 4: Hold Pre-Job (Planning) Kickoff Meeting with Internal Team Members to Assign Responsibilities

- Arrange a meeting time and location, and ensure that enough time has been allocated to assign planning responsibilities.
- Invite internal team members who will be involved during the construction execution process, including project manager, field supervisor, accounting, purchasing, and the director of field operations.
- Follow and complete a checklist that describes the planning activities that must be completed and who is responsible for completing them. Figure 1.6 shows a sample planning kickoff meeting agenda.

Scope and Contract Review

A scope review involves reviewing all contract documents that will be used to execute the work and produce a high-quality end product. It involves becoming familiar with the detailed project requirements, including the systems that will be installed, equipment that must be powered, the materials that are specified, and the labor that will be required to complete the job successfully. A contract review involves reading the entire contract agreement, including general and special conditions, to identify any unusual or special provisions that must be accomplished as part of the scope of work. The scope and contract review process consists of eight activities.

Activity 5: Review Contract for Unfavorable or High-Risk Clauses

- Complete the contract review checklist shown in Figure 1.7.
- Identify all required bonds, permits, certificates, and insurance requirements. A specialty contractor might be required to provide a performance and payment bond or a completion bond. The specialty contractor might also be responsible for obtaining specialty permits and must provide a certificate of insurance demonstrating that they have the insurance required by the contract.

1. Project overview, including:
   a. Project name
   b. Location
   c. Type of work
   d. Contract cost
   e. General scope of work
   f. Identify potential opportunities and challenges of the project

2. Introduce internal team members, including:
   a. Project manager
   b. Field supervisor/Foreman
   c. Estimator
   d. Accounting representative
   e. Other internal team members (i.e., CAD operator, purchasing agent)

3. Identify external team members, including:
   a. Owner/Customer
   b. Architect/Engineer
   c. General contractor/Construction manager
   d. Other specialty contractors
   e. Vendors/Suppliers

4. Review the general scope of work
   a. Provide an overview of the entire project scope of work and your specific scope
   b. Review major work performed by others

5. Review the contract cost
   a. Review direct costs and the contract cost
   b. Review overhead, profit, and contingency funds

6. Review purchasing of materials, equipment, and services from subcontractors/vendors
   a. Review subcontractors and suppliers scope of work and qualifications
   b. Identify all pre-contract commitments and promises
   c. Verify minority business requirements

Figure 1.6 Pre-job planning kickoff meeting agenda

- Review contractual billing requirements. The owner typically will require that pay requests be submitted by a date identified in the contract. Furthermore, the general contractor's own accounting department will often identify a date they would like to receive both draft and final pay requests. These dates should be identified upon initiation of the project to ensure timely payment is received.
- Review and understand procedures for requesting change orders. Review the contract and specifications carefully to understand the procedures that must be followed to request a change order and recover the additional time or money spent on work items. It is strongly recommended that you set up a change management process and ensure all team members understand and follow the process.

7. Review the schedule and milestones
   a. Review the owner-furnished schedule
   b. Review any preliminary schedule submitted with the bid or construction schedule
   c. Review work by others that will impact your work
   d. Identify important material and equipment delivery dates

8. Review manpower requirements and labor rates
   a. Review the manpower loading chart
   b. Identify the estimated crew mix
   c. Identify any wage increases

9. Review the contract and identify any special clauses that require careful consideration

10. Review administrative procedures, including:
    a. Administrative setup
    b. RFI procedures and setup
    c. Submittal process and setting up the submittal tracking system
    d. Change order procedures and setting up the change management system
    e. Field reporting requirements
    f. Budget preparation and billing procedures
    g. Cost control and setting up the tracking system

11. Review special safety issues

12. Review site logistics and material storage and staging, including:
    a. Site access
    b. Parking
    c. Material delivery and movement procedures
    d. Material storage locations
    e. Trailer locations
    f. Site cleanup requirements
    g. Temporary power and lighting requirements

13. Review bond, permit, and certificate of insurance requirements

14. Review other items specific to this project

Figure 1.6 *Continued*

- Identify whether disputes must be resolved through alternative dispute resolution (ADR), such as mediation or arbitration. ADR requirements are typically covered in the contract or specifications. Be sure to understand whether ADR is contractually required and whether it is binding or nonbinding.

Activity 6: Project Manager Reviews Plans, Specifications, and Schedule

- Order extra sets of plans and specifications if necessary. Be sure to order enough plans so that there is at least one set in the main office (project manager's copy), two sets in the field trailer (field supervisor's

Completed	Completion date	Item
☐		Check for ambiguous clauses and seek clarification
☐		Check for one-sided clauses that favor the other party
☐		Cross-reference clauses to thoroughly understand the meaning of clauses
☐		Identify discrepancies and conflicting clauses and seek clarification
☐		Check for agreement between plans and specifications and seek clarification if there is any discrepancy
☐		Identify any "killer clauses" that assign you full responsibility for everything
☐		Determine whether the contract clearly favors the other party (owner/contract manager/general contractor) and seek more equitable contract terms
☐		Review the indemnification clause and ensure your insurance coverage can sufficiently cover the risk
☐		Evaluate the coordination clause to determine each party's responsibility to coordinate the work
☐		Determine your rights and responsibilities if the sequence of work is changed or if out-of-sequence work impacts the project
☐		Carefully review all time-sensitive clauses that cover commencement, completion, milestones, accelerations, delays, and progress
☐		Review and understand your rights to receive damages if the schedule is accelerated or delayed by other parties
☐		Determine whether there is a no-damages-for-delay clause and review your rights and responsibilities
☐		Review and understand the clause that describes waivers
☐		Review the concealed conditions clause and identify the circumstances that will entitle you to compensation
☐		Review the force majeure clause and identify the circumstances that will entitle you to a time extension
☐		Evaluate clauses that identify procedures for seeking change orders
☐		Evaluate clauses that identify the compensation for extra work
☐		Identify payment provisions, especially a "pay when paid" clause and a "pay if paid" clause

Figure 1.7 Contract review checklist

copy and the marked up copy), and one or more copies that can be distributed to foremen.

- Complete the scope and schedule review checklist shown in Figure 1.8.
- Compare the scope identified in the contract to the scope from the bid submission to verify that they match. The bid submission letter

Completed	Completion date	Completed by	Item
☐			Contract
☐			Plans
☐			Specifications
☐			Cost estimate and bid breakdown
☐			Referenced/Applicable codes or regulations
☐			Quality requirements
☐			Safety requirements
☐			Special conditions
☐			Addendums
☐			Temporary power and lighting requirements
☐			Owner/Contract manager/General contractor-furnished materials
☐			Vendor pricing and qualifications
☐			Owner/Contract manager/General contractor schedule
☐			Internal schedule submitted with bid
☐			Work sequence and work by others
☐			Required coordination with other trades
☐			Material and equipment deliveries
☐			Anticipated weather problems or holidays
☐			Required labor and crew mix
☐			Labor rates and potential increases
☐			Administrative procedures (submittals/RFIs/changes)
☐			Access doors
☐			Asbestos abatement
☐			Cleanup
☐			Crane
☐			Demolition and removal
☐			Electric motors
☐			Excavation and backfill
☐			Hoists for personnel
☐			Hoists for materials
☐			Interior layout
☐			Painting
☐			Scaffolding
☐			Site access
☐			Site surveying
☐			Underground utilities

Figure 1.8 Scope and schedule review checklist

will typically identify the scope of work, clarifications, and inclusions/exclusions. This letter should be compared to the scope identified in the contract to determine if there is a discrepancy. Any discrepancy should be discussed with the estimator and company officer to determine how to resolve it.

- Compare the plans to the specifications to identify any discrepancies. There are many ways that the plans can differ from the specifications. In general, the specifications will supersede the plans.

- Review the schedule provided by the general contractor to determine whether your work can be completed within the contractual time frame. Plan the work strategy based on this finding. Occasionally, the time required to complete your work may be longer than the contractual duration. In such cases you should plan from the beginning for overtime, overmanning, or acceleration. The need to accelerate should, ideally, be established during the bidding stage, but the techniques for accelerating the work will be selected after award.

Activity 7: Field Supervisor Reviews Plans, Specifications, and Schedule

- Complete the scope and schedule review checklist shown in Figure 1.8.
- Review best field practices or lessons learned. If the company has a book of best field practices or lessons learned, the field supervisor should review this book before planning the execution of the new project to avoid reinventing the wheel. If no best practices or lessons learned exist, the field supervisor can talk to other field personnel or hold a meeting to identify the best methods for executing the new project.
- Identify value engineering (VE) or prefabrication opportunities (see Activity 11). While reviewing the plans and specifications, and after reviewing best practices and lessons learned, the field supervisor should begin assembling a list of installation methods, materials, or prefabrication of systems that could result in a savings to the contractor. Note whether the technique requires owner or engineer approval and whether the savings must be shared with the owner.
- Identify labor requirements and begin selecting foremen. The field supervisor should identify the number of workers and foremen that will be required each week of the project. This can be determined by creating a manpower loading chart. The field supervisor can select foremen to cover specific systems or specific areas such as one foreman per floor on a multistory building. Foremen can also be selected based on skills, including experience working on a hospital or a power plant.

- Identify special tools that will need to be purchased or assembled. Along with the list of potential cost-saving changes that are identified while reviewing the plans and specifications, the field supervisor should also assemble a list of special tools that might be needed to install the materials identified in the contract documents or to complete the installation more efficiently.

Activity 8: Create a List of Issues That Need to Be Resolved and Begin the Request for Information Process

- Establish a Request for Information (RFI) process (see Activity 16) and create a list of questions that require clarification. While reviewing the plans and specifications, the field supervisor should also assemble a list of discrepancies and issues that need to be resolved in order to execute the work efficiently. This list will kick off the RFI process and should be resolved as quickly as possible. You should consider holding a special meeting to get these issues resolved.
- Formalize the RFIs by assigning them a trackable number and submitting them to the owner/contract manager/general contractor. Activity 16, which is an administrative procedure that involves initiating an RFI process, is closely related to Activity 8. Most companies have a formal RFI process that includes a hierarchy of personnel who must review the RFI and the standard submission format, such as emailed, faxed, or mailed RFI form with a clear routing sequence. All RFIs should have a unique number that permits them to be tracked for resolution.

Activity 9: Conduct Site Visit

- Complete the site visit checklist shown in Figure 1.9.
- Examine site access and layout, including parking, material delivery points, and material lay-down and storage. Site logistics is especially important on small, bounded sites. You should determine whether: (1) workers can park on the site or must park in a nearby garage for a fee; (2) materials can be delivered at any time or only at certain hours of the day; (3) a material and/or office trailer will be permitted on the site; and (4) materials can be stored on the site under secure conditions.
- Identify locations and availability of material and personnel lifts, elevators, cranes, scaffolding, and forklifts. Verify whether lift, crane, and machinery time must be scheduled in advance. Also determine

Completed	Item	Notes
☐	Access into and out of the site	
☐	Circulation throughout the site	
☐	Material and equipment delivery routes	
☐	Material storage and staging locations	
☐	Office trailer or office space	
☐	Temporary power and lighting locations	
☐	Existing underground utilities location	
☐	Existing aboveground utility locations	
☐	Location of existing interior systems	
☐	Progress of the demolition	
☐	Progress of the site work	
☐	Progress of the site layout/Surveying	
☐	Asbestos abatement has been completed	
☐	Work completed to date	
☐	Presence/Location of the crane	
☐	Presence/Location of the personnel lift	
☐	Presence/Location of the materials lift	
☐	Potential coordination issues with others	
☐	Anticipated weather problems	
☐	Housekeeping conditions	
☐	Special site conditions	
☐	Safety issues or concerns	

Figure 1.9 Site visit checklist

whether you must supply your own scaffolding or whether it will be furnished by the general contractor.

- Create a plan for installing temporary power. While on the site visit, sketch out a plan for installing temporary power and lighting and determine how these systems will be maintained. A carefully thought-out temporary power and lighting plan can prevent spending more than was estimated to install and operate these systems.

- Verify existing conditions and compare them to the conditions shown on the plans and described in the specifications. If the site contains facilities that must be renovated or demolished, or if the site has existing structures or utilities, a site visit should be conducted to compare the information on the plans to the actual conditions on the site. If a building is being renovated, the plans will show the locations of all existing systems (mechanical, electrical, plumbing, and so on). These locations should be verified and marked up on the plans to avoid conflicts with coordination later.

Activity 10: Compare Estimated (Bid) Work Activities and Materials to Planned Performance

- Compare the estimator's concept of how to perform the work to typical field operations and document the differences. An estimator typically has less time to plan out the work sequence, materials, and methods than the project manager and field supervisor. As a result, the project manager and field supervisor should compare how they plan to perform each work item to how the estimator estimated the same work item. For example, the estimator may not have planned to perform any prefabrication of systems, but the project manager might identify several work items that could be prefabricated prior to installation.
- Discuss unclear methods or discrepancies with the estimator. If the project manager or field supervisor is unclear about how the estimator planned a particular work item, they should seek clarification to make sure they do not unnecessarily expend time, money, or work hours.
- Determine the cost difference between the as-bid and planned performance of the work. If a less expensive work method can be identified, the project manager should document the cost savings in the cost control report so that the additional funds can be added to profit or cover cost overruns.

Activity 11: Identify Value Engineering and Prefabrication Opportunities and How to Simplify the Work

- Review VE or prefabrication opportunities that were identified in the turnover meeting between the estimator and project manager.
- Review procedures for formally requesting VE consideration, if required. Some VE changes involve substituting materials or modifying installation routes, which might require approval by the owner or architect. Other VE changes may be strictly associated with methods and might not require approval. If approval is required, the project manager should understand the timeline and procedures for receiving approval to determine whether the VE suggestion will be feasible.
- Search for and identify additional VE and prefabrication items. In addition to the VE and prefabrication opportunities that were identified during the turnover process, the project manager and field supervisor should seek additional opportunities that would result in a time or cost savings. Opportunities that do not require customer approval are particularly appealing.

- Identify additional ways to simplify the work. Simplification might involve ordering materials in bulk so that they are readily available on the site, or simplification might involve marking materials with tags to identify items. There are numerous ways that the field supervisor and foremen can simplify the installation process, and these methods should be explored during the scope review process.
- Price out the cost difference between the as-bid and VE options. This is an exercise performed by the project manager to determine the cost savings or cost expenditure to implement a VE option. Although the goal is typically cost savings, occasionally a VE option will be selected to save time.

Activity 12: Prepare Construction Takeoff

- Take off the materials, equipment, and systems in the order they will be constructed. The construction takeoff is often performed to: (1) verify that the bid quantities are correct; (2) calculate accurate quantities for ordering materials and equipment; (3) begin sequencing the work; and (4) validate the cost estimate. It is an important exercise that can identify problems and opportunities early in the planning process.
- Code each plan sheet as you take it off so that you can return to it later and immediately identify the quantity of various materials shown on the sheet. A common method to prepare a quantity takeoff is to color-code the plan sheets for easy reference. Each system should be coded individually, and a legend should be provided.
- The final quantities should be identified by the units in which they will be purchased. For example, if a material such as wire will be purchased by reel, then the number of required reels should be identified rather than the linear feet of wire. The construction takeoff should be the first step to purchasing the actual materials needed to complete the work.
- All assumptions should be noted on the quantity takeoff sheets.
- Be sure the quantities that are estimated include a waste factor. An accurate quantity takeoff will provide a definitive amount of material needed to complete the work. However, since most materials will need to be cut or bent to fit its exact placement in a system, a waste factor should be added to the quantity.
- Compare the construction takeoff to the bid takeoff to identify significant differences or discrepancies. It is unlikely that the bid estimate and the construction takeoff will match exactly. However, most quantities

of materials and labor should be within five percent of the bid amount. Significant differences might be an indication of a bid mistake or a misunderstanding of how the work will be performed. Any major discrepancies should be reviewed by the project manager and estimator.

Administrative Setup

The administrative setup involves creating standardized paper and computer files that will be used to manage correspondence, changes, submittals, schedules, and progress. A standardized administrative setup process is necessary to run an organized and efficient project. The administrative setup process consists of six activities.

Activity 13: Set Up Project Files and Create Contact List

- Use the file system checklist shown in Figure 1.10 to create paper files.
- Create a contact sheet that lists all team members and their company contact information. The list should include owner, architect, engineer, and general contractor contacts, and might also include consultants, facility occupants, maintenance staff, city inspectors, and all internal team members (project manager, field supervisor, foremen, administrative assistant, and so on).

Completed	File number	File description
☐		Project information and contacts
☐		Cost estimate and bid submission
☐		Contract agreement
☐		Contract documents
☐		Budget and pay requests
☐		Purchase orders
☐		Subcontracts
☐		Materials folders
☐		RFIs
☐		Submittals
☐		Change orders—pending
☐		Change orders—approved
☐		Correspondence
☐		Meeting minutes
☐		Daily/Weekly field report
☐		Progress reports

Figure 1.10 File system checklist

*Activity 14: Set Up Computerized Tracking and Control System
(Forms, Database, Schedule, Tracking)*

- Verify that the accounting office has assigned a project number and entered initial information in the cost control system. During the bidding stage, or upon notification of award, the estimator may have requested that the accounting office enter the project into the accounting database. If no project number has been assigned, the project manager should request a number from the accounting department.
- Use the file system checklist shown in Figure 1.10 to create computer files.
- If a separate project management software system is used, set up the project in the system. Commercially stand-alone project management software is available, as well as web-based project management systems. Most software can be tailored to the needs of the company. The software should be tailored to reflect the processes established by the company. If commercial software is not used, the project manager can set up the files in a word processing and spreadsheet program.
- If a separate computer scheduling system is used, set up the schedule in the system (see Activity 37). Prior to entering activities in the schedule, the project manager and field supervisor should develop a sequence of work.

Activity 15: Initiate a Change Management System

- Review the contract to identify required change order, field change, and time-and-materials procedures. If the contract does not specifically address change orders, field changes, and general change procedures, the project manager or a company officer should request a separate meeting to discuss the company standard procedures.
- Review your company's standard procedures for initiating, requesting, and processing change orders and field changes.
- Develop a log, with sequential numbering, to track all changes, including change orders, field changes, and time-and-materials requests as shown in Figure 1.11.

*Activity 16: Initiate a Request for Information (RFI) Tracking and
Processing System*

- Develop a log with sequential numbering to track all RFIs as shown in Figure 1.12. Many of the commercial project management software programs will provide a method for initiating and tracking RFIs.

C.O. number	Date submitted	Description of change	Initiated by	Authorized by	Associated RFI number	C.O. amount	Status

Figure 1.11 Change order log

RFI number	Date submitted	Description of information needed	Submitted to	Date of response	Initiated by	Status

Figure 1.12 Request for information log

- Determine whether RFIs will be submitted by email, fax, or postal mail.
- Review your company's standard procedures for processing RFIs and use a company standard form for submitting RFIs.
- Ensure each RFI also identifies a proposed solution and price change.

Activity 17: Initiate a Submittal Tracking and Processing System

- Develop a log with sequential numbering to track all submittals, including those of vendors and subcontractors as shown in Figure 1.13. Review the specifications to determine whether the customer has included a submittal log. If so, this information can be transferred to your own company tracking system. If not, use either commercial project management software or create a spreadsheet. Go through each section of the specifications to determine whether it identifies a submittal that you or your suppliers or subcontractors are responsible for providing. Enter this information in your submittal tracking log.
- Review your company's standard procedures for processing submittals and use a company standard form for submissions.
- Verify that each submittal processing form identifies a response date, which is associated with timely ordering and delivery of the materials and equipment. Be sure to review the work sequence and schedule carefully to identify when each material or equipment item will be installed. Then, annotate on the submittal form the date you will need an approval to order and receive the materials or equipment on time to avoid a delay. This technique also puts the architect/engineer on notice in case there is a delay that is disputed later.

Activity 18: Develop a Labor Requirements/Expectations Letter

- For projects that have special requirements, such as drug testing, background check, or special safety training, develop an expectations letter that must be reviewed and signed by crew members.
- For projects that will require hiring workers from the union hall, develop an expectations letter that must be reviewed and signed by crew members.

Buyout Process

The buyout process involves purchasing all materials and equipment—and hiring all subcontractors—that will be necessary to complete the project. Consequently,

CSI section	Date submitted	Revision number	Submitted item	Supplier	Date to arch/eng	Date approved	Date to supplier	Order date	Delivery date

Figure 1.13 Submittal log

it involves evaluating material pricing, reviewing the qualifications of vendors and subcontractors, selecting the successful parties, issuing purchase orders, and initiating the submittal process. The buyout process consists of six activities. It should be noted that Activity 24 (Submittal Processing and Tracking) should be conducted in conjunction with Activity 17 (Submittal Tracking and Processing System).

Activity 19: Request and/or Review Subcontractor/Supplier/Vendor Prices and Qualifications

- Request subcontractor/supplier/vendor pricing if it was not requested or received during the bidding stage. Two scenarios are possible. In the first scenario, actual pricing was requested and received from interested subcontractors/suppliers/vendors during the bidding stage. In the second scenario, estimated pricing was obtained from a cost database. If actual pricing was quoted, you may want to request "best and final" pricing (also called "buy pricing") from all interested subcontractors/suppliers/vendors. If estimated pricing was obtained from a database, you will want to request actual quotations.
- Compare actual subcontractor/supplier/vendor scope of work with the scope identified in the subcontractor/supplier/vendor bid submission. The subcontractor/supplier/vendor scope of work should be discussed when the quotation is requested. Ideally, the project manager will hold a meeting to review the scope with each potential subcontractor/supplier/vendor. Then, after quotations have been received, the project manager should compare the submitted scope of work with the actual scope of work to identify any discrepancies. These differences will need to be resolved prior to making an award.
- Evaluate the subcontractor/supplier/vendor qualifications. The project manager should evaluate whether: (1) the subcontractor/supplier/vendor has performed this type (and/or size) of work before; (2) you have worked with the subcontractor/supplier/vendor before; and (3) the subcontractor/supplier/vendor performance has been satisfactory on previous jobs.
- Compare pricing and qualifications among the subcontractors/suppliers/vendors. Use a spreadsheet (or commercial project management software) to create a form to compare price and qualifications of the subcontractors/suppliers/vendors as shown in Figure 1.14.

Item type: _____		Vendors			
Review items:	**Budget**	**Vendor A**	**Vendor B**	**Vendor C**	**Vendor D**
Was the item bid per plans and specs? [Yes/No]					
Was the item bid per the scope of work? [Yes/No]					
Are the workers union or nonunion?					
How long will it take to order and deliver the item?					
List any exclusions					
Was tax included in the bid price? [Yes/No]					
Did the vendor acknowledge all addenda? [Yes/No]					
Base bid price					
List any alternate pricing provided					
List adjusted bid price if alternates accepted					

Figure 1.14 Subcontractor/supplier/vendor comparison spreadsheet

Activity 20: Negotiate Pricing and Contract Conditions and Select
Subcontractors/Suppliers/Vendors

- Discuss possible cost savings with potential subcontractors/suppliers/ vendors. Many suppliers can offer significant discounts if materials are purchased in bulk rather than weekly. Furthermore, suppliers and subcontractors may be able to offer VE ideas, such as material substitutions or innovative purchasing techniques. These savings should be discussed with potential subcontractors/suppliers/vendors before making a final selection.
- Review the contract or purchase order terms and conditions with potential subcontractors/suppliers/vendors prior to award. Specifically discuss (1) shipping and delivery terms, especially if special delivery or packaging is required, (2) who will conduct an inventory when the materials are delivered, (3) documenting, returning, and replacing damaged materials, (4) consequences of early or late deliveries, (5) signing authority, and (6) other terms specific to each project.
- Select all successful subcontractors/suppliers/vendors and issue a letter of intent. After carefully reviewing pricing and qualifications of all subcontractors/suppliers/vendors, select the successful parties. These companies should be notified by phone, and a letter of intent should be issued (by fax) if there will be a delay in processing the contract or purchase order. The letter of intent will authorize the contractors/ suppliers/vendors to purchase materials or begin work on the project while the official paperwork is being processed.

Activity 21: Develop and Issue Purchase Orders and Contracts
for Materials and Equipment

- Develop the purchase orders or the contracts for subcontractors/ suppliers/vendors. Standard subcontracts, such as American Institute of Architects (AIA) or Associated General Contractors of America (AGC) standard forms, should be used in place of a purchase order for work that involves materials and labor because the subcontract will outline responsibilities, liability, insurance, and indemnification. The purchase orders are typically prepared by the accounting department or by the project manager with the assistance of the accounting department.
- Process and issue purchase orders or subcontracts. Ensure that a company officer from both your company and the subcontractor/supplier/ vendor has signed the purchase order or subcontract to acknowledge

the terms and conditions. The purchase order or subcontract should also note whether the materials, equipment, or services are to be ordered or supplied immediately or whether the supplier should hold the order until an order and delivery date has been established.

Activity 22: Order Long-Lead-Time Materials and Equipment

- Compare the sequence, schedule, and materials/equipment to identify long-lead-time items. When reviewing the scope of work, the project manager and/or field supervisor should document those materials and equipment that might have a long lead time or might impact the schedule if delivery is delayed. If these items were not annotated during scope review, the project manager and/or field supervisor will need to carefully re-review the scope and schedule to identify long-lead-time items that could impact work progress.
- Negotiate and issue purchase orders and subcontracts for long-lead-time items before negotiating standard purchase orders and subcontracts. If a material or equipment item might require a long lead time before it can be delivered, then the purchase of the item should be negotiated and authorized before non-long-lead-time items are evaluated and purchased to ensure timely ordering and delivery.
- Issue a letter of intent or purchase order immediately to release long-lead-time items for order and delivery. Often, long-lead-time material and equipment pricing is negotiated prior to the bid submission. Then, after the contractor receives the notification of a pending award, the project manager can immediately issue a letter of intent or purchase order requesting that the item(s) be ordered. The letter of intent or purchase order can be faxed, with a hard copy mailed the same day, to expedite the ordering process.

Activity 23: Request Submittals, Cut Sheets, and Shop Drawings

- Issue the contract document, and then the subcontractor/supplier/vendor should be requested to assemble and submit samples, cut sheets, or shop drawings. As part of the contract agreement, the subcontractors/suppliers/vendors should be requested to submit items as identified in the specifications. The project manager should identify, in the contract document, which submittals are required, and these items should be verified with the subcontractors/suppliers/vendors. Occasionally, submittals are requested during the bidding stage as part of the subcontractor/supplier/vendor bid submission, and some cut sheets are available through a cost database. Therefore, these submittals might already be available for processing.

- Identify a deadline by which the submittals, cut sheets, and shop drawings must be submitted to your company. Two submittal strategies are possible. In the first strategy, the contractor might be required to assemble a binder with all of the required submittals in it so that all submittals can be processed at the same time. In the second strategy, the contractor might be able to submit items in the order of their installation or as they are received from subcontractors/suppliers/vendors. In either case, the project manager should review the schedule and identify a deadline for receiving each submittal so that it can be processed in a timely manner to avoid delays. This deadline should ideally be identified in the contract document (purchase order or subcontract).

Activity 24: Develop and Process Log and Book of Submittals, Cut Sheets, and Shop Drawings

- Develop a log with sequential numbering to track all submittals, including those of vendors and subcontractors (see Activity 17 and Figure 1.13). Review the specifications to determine whether the customer has included a submittal log. If so, this information can be transferred to your own company tracking system. If not, use either commercial project management software or create a spreadsheet. Go through each section of the specifications to determine whether it identifies a submittal that you or your suppliers or subcontractors are responsible for providing. Enter this information in your submittal tracking log.
- Assemble two or more binders of all draft and approved submittals. At least two complete binders of submittals should be assembled so that the project manager has a copy in the office and the field supervisor has one on the jobsite. If a complete binder of submittals will be processed for approval by the architect/engineer, additional copies of the binder will be required so that the architect/engineer and the owner each receive a copy.
- Submit and track required submittal items. Use the submittal log (Figure 1.13) to track receipt and approval of all submittals. Issue dunning letters, as necessary, if the architect/engineer does not approve or return the submitted items by the requested deadline.

Material Handling Plan

Material handling planning involves establishing processes for ordering, receiving, staging, and storing major materials and equipment on the jobsite or at a storage location. As part of the material handling planning process, participants should also review general site logistics to ensure the materials and site facilities are located efficiently and promote maximum productivity of workers and equipment. Material handling planning consists of two activities.

Activity 25: Develop Material Delivery and Handling Plan

- Review the following material delivery/storage and site logistics best practices:
    - Assign one person the responsibility of managing material and equipment delivery, handling storage, and staging
    - Establish a standard unloading crew that consists of laborers or apprentices
    - Establish standard procedures for receiving, handling, and storing materials and strictly enforce the standard
    - Develop a storage site layout that identifies where the materials and equipment are stored and annotate the location on the material and equipment delivery and storage log
    - Develop a storage site identification system that provides a method to document and track the location of all materials that have been delivered to the jobsite
    - Establish storage space for each major material item or group
    - Sort and store the materials as soon as they are delivered to the site
    - Allow storage space for waste or excess materials and remove them as soon as possible
    - Ensure the material and equipment are adequately secured and protected from the elements
    - Make sure that the materials manager is notified of any pending deliveries
    - Return all damaged, excess, or incorrect materials to the vendor immediately to keep the site free from clutter
    - Arrange to have materials delivered just before you need them so that storage and handling are kept to a minimum
    - If materials are ordered in bulk, ask the vendor to store the materials at their office until you need them on the jobsite (also consider paying extra for this option)

- Consider using a material consignment trailer, where the vendor inventories the trailer each week, restocks it, and only charges you for the materials you use
- Try to place materials/equipment at the location where they will be used to improve access to the materials/equipment and minimize handling
- Arrange to have materials packaged for efficient unloading, handling, and installation
- Determine ahead of time what equipment will be needed to unload and handle material deliveries
- Evaluate the capacity of material lifts, freight elevators, and cranes to ensure they can safely move the materials and equipment
- Locate your toilet facilities as close to the work areas as possible
- Locate your break facilities and trash containers as close to the work areas as possible
- Develop a map of facility locations and distribute it to workers and suppliers/vendors

- Establish and maintain a file of delivery receipts and packing slips. The person who is in charge of receiving materials and equipment should maintain the file of delivery receipts and packing slips. Other individuals who might receive shipments (in the main office or at the site office) should ensure that the material/equipment delivery manager ultimately receives the receipts so that accurate records can be maintained.
- Establish a material and equipment delivery and storage log. Figure 1.15 provides an example of a material and equipment delivery and storage log. The log should be maintained by the person in charge of receiving materials and equipment.
- Create a material and equipment delivery schedule. During the scope and schedule review (or the buyout process), the project manager and field supervisor should annotate when materials and equipment will be ordered and delivered to the site. These annotations should be developed into a formal schedule of deliveries. This schedule should be distributed to the material manager, and then the schedule should be reviewed and updated regularly by the material manager and field supervisor.
- Establish standard procedures for receiving, handling, and storage of materials and equipment. The formal procedures should document delivery hours, site entry and exit points, site circulation, and procedures for delivering materials to the work location or storage area. The plan should outline who will determine the unloading point (work

Date received	P.O. number	Description of item received	Quantity	Received by	Storage location	Damaged items	Notes

Figure 1.15 Material and equipment delivery and storage log

location or storage location) and how this information will be conveyed to the vendor. The plan should also identify authorized signing agents and the name and contact information of the material manager and field supervisor. The plan should also identify how to annotate delivery, inspection, acceptance, and storage location of all materials and equipment delivered to the site. The procedures should be strictly enforced to maximize the efficient and orderly receipt of materials.

Activity 26: Develop Material Storage and Staging Plan

- Complete the site logistics review checklist shown in Figure 1.16.
- Review the material delivery/storage and site logistics best practices.

Completed	Completion date	Item
☐		Review the site layout and identify placement of materials and facilities to maximize productivity
☐		Identify site entry and exit points and plan possible vehicle circulation
☐		Identify procedures for receiving materials and authorized personnel
☐		Identify material storage locations
☐		Determine and document equipment that will be needed to unload and move materials (e.g., cranes, forklift, pallet jacks)
☐		Evaluate material lifts, freight elevators, and cranes to determine whether they can support the size and weight of the material items
☐		Determine and order special tools associated with material handling (e.g., box cutters, bar code readers, computers)
☐		Establish a receiving crew that consists of laborers or apprentices who will unload trucks and move materials
☐		Select a worker to be in charge of material handling, including inspection and inventory of delivered items
☐		Establish standard procedures for receiving, logging, handling, and storing materials and equipment on the jobsite or at an offsite location
☐		Establish delivery dates for all materials and develop a schedule of deliveries
☐		Evaluate purchasing options to ensure materials are ordered to promote efficient unloading, storage, and installation
☐		Evaluate the benefits and pitfalls of prefabrication in terms of delivery, storage, handling, and installation
☐		Review the material delivery, handling, storage, and staging best practices

Figure 1.16 Site logistics review checklist

- Develop a storage site layout that identifies where the materials and equipment are stored. Show the locations of all major groups of materials or equipment as well as consolidated storage locations. An identification system should be developed so that materials and equipment can be logged in and tracked by their storage location identification.

Budget Preparation

Budget preparation involves developing a cost code scheme and breaking down materials, labor, and overhead into discrete categories that can then be used for billing during execution. As part of the budgeting process, a breakdown of costs for tracking progress must be performed, and this breakdown may be different from the cost breakdown identified on a schedule of values used for billing. Developing a budget (for tracking) and schedule of values are the key elements of cost and cash flow management. The budget preparation process consists of three activities.

Activity 27: Develop, Review, or Expand Cost Code Scheme

- Decide whether to use a cost code scheme based on the Construction Specifications Institute (CSI) MasterFormat.
- Review the cost estimate to identify the existing cost code breakdown, and decide whether additional breakdown is necessary. During the preparation of the cost estimate, it is likely that the estimator developed an initial cost breakdown based on the standard company cost codes. This breakdown should be reviewed to identify whether additional codes will be needed to track and bill the work.
- Review the company standard list of cost codes or use the CSI MasterFormat and decide which codes will be needed for breaking down the work for tracking and billing. A cost code scheme can provide an outline for breaking down materials, labor, and equipment. Keep in mind that this breakdown should reflect how the work will be performed, and it should be as accurate and detailed as practical for tracking and billing.
- Add new codes for work items that are not on the standard list. The standard list will typically contain only standard items encountered on most projects. If your job, for example, calls for installing stadium lighting that must be lifted by a crane, you probably will not have the proper line items to cover this specialty work. Therefore, you will need to work with your accounting department to temporarily create new codes that cover the specialty items on your project.

- Assign additional codes to specify the costs associated with material, equipment, labor, subcontractors, and other miscellaneous costs. Each line item or task is usually divided into material costs and labor costs. An example for making such a division might be to add a dash after the cost code and then add the additional code. For example, if the cost code for cable is 260500 and the additional code for material is M and for labor is L, then the code for cable (material) would be 260500-M and for cable (labor) would be 260500-L.

Activity 28: Develop Budget by Breaking down Labor, Material, Overhead, and Profit Costs

- Create the budget in concert with the labor and materials tracking report. One of the main purposes of a budget is to track material and labor costs. Therefore, you should decide which cost and work items you want to track, and develop your budget to match the items that will be tracked. Figure 1.17 presents a sample budget breakdown and tracking form.
- Establish a budget with sufficient line items to identify potential problems yet simple enough to avoid time-consuming data entry. Remember that the goal of material and labor cost tracking is to make sure the project cost is not exceeding the estimated amount. Therefore, make sure there are enough line items to pinpoint a specific problem work item. However, the budget should not be so detailed that it requires an unbalanced proportion of time to enter all of the cost data.
- Verify that labor and materials can, and will, be reported according to the budget line items. The true test of a good budget is to identify whether the field can track labor against the line items. The field supervisor should be able to tell you if the budget is too detailed and whether it would require too much time to track labor against all of the line items. If the budget appears to be too detailed, consider rolling up or combining several line items into summary items.

Activity 29: Develop Schedule of Values

- Review the contract to identify the contractual format and process for developing a schedule of values. Some contracts specify a standard AIA form be used to submit the schedule of values for approval. If a standard AIA form is not specified, prepare a company schedule of values as shown in Figure 1.18.

CSI division	Description	Labor	Materials	Equipment	Subcontracts	Overhead and profit	Total

Figure 1.17 Budget breakdown and tracking spreadsheet

Item (A)	Description (B)	Contract value (C)	Previously completed (D)	Completed this period (E)	Completed to date (F) = (D)+(E)	Percent complete (G) = (F)/(C)	Balance to finish (H) = (C)−(F)	Retainage (I)

Figure 1.18 Company schedule of values

- Create the schedule of values in concert with the billing process. Review the company billing procedures. Then, while preparing the schedule of values, think about how easy or difficult it will be to determine the percent complete of each of the line items. If the process is difficult, you might need to add more details to accurately capture the work that has been completed.
- Consider rolling up the budget, so that the schedule of values has the same summary line items but fewer sub-line items. One way to develop the schedule of values is to copy the budget but eliminate the detailed line items. Instead, the summary line items are used.

Layout and Sequencing Plan

Layout and sequencing planning is the process of developing a sequence of work, laying out that sequence in a series of drawings for field execution, and developing installation instructions for crew members. As part of the layout and sequencing process, CAD installation drawings may be developed and distributed to foremen and field crew to minimize questions and maximize productivity of the workforce. The layout and sequencing planning process consists of three activities.

Activity 30: Develop Installation Sequence and Layout Drawings

- Organize the project by areas, by floors, or by systems. Breaking down the project in areas or systems has several benefits. For example, separate foremen can be assigned to manage each area/system, providing a smaller span of control. Furthermore, the schedule and tracking programs can be developed to match the area/system breakdown.
- Allocate sufficient time to mentally think through the sequencing of all work processes from start through completion. The project manager, field supervisor, and key foremen might participate in this mental exercise, which might be conducted as a brainstorming meeting. One participant should be assigned the responsibility of taking minutes so that the final sequencing can be transferred to layout and installation drawings. Key participants should identify every activity that must be accomplished to develop a completed product. Initially, all ideas should be written on a white board or flip pad, and then the activities can be sequenced, eliminated, or consolidated through a series of iterations.
- Determine whether to create sequence and layout drawings by hand (marked-up drawings) or by creating new CAD drawings. This decision might be driven by company capabilities. Many companies are hiring CAD operators to work on the jobsite creating layout and

installation CAD drawings. These drawings are handed out each morning to foremen who direct the activities of their crew members. If the company does not have CAD capabilities, hand sketches or marked-up plan sheets can also be distributed to foremen as installation drawings. Both techniques appear to minimize questions, improve productivity, and contribute to a more organized and efficient installation process.

- Create daily installation drawings at least one day prior to the date when the work needs to be performed so that the drawings can be distributed each morning to foremen and field crews. The person preparing the installation drawings should stay one or more days ahead of the installation schedule. Ideally, at least a week's drawings should be prepared one week or more ahead of the schedule.

- Review the sequence daily or weekly to ensure the project is progressing as expected. The sequencing and installation plans should be used in conjunction with the schedule to ensure the project is progressing as planned. Any deviations from the planned sequence or schedule should be discussed with the general contractor and documented. Some flexibility with the sequencing or installation of the work might be required to keep the project progressing as a whole.

Activity 31: Develop Field Instructions

- Review the sequence and installation process to identify any additional information that might minimize questions and improve productivity.

- Develop field instructions and drawings for repetitive work when repetitive work is scheduled on a project. A good example of a project that will include repetitive work is a hotel or dormitory. Each room is similar or identical to all of the other rooms. Therefore, a single installation sequence, drawings, and set of instructions can be developed for the crew performing the repetitive work. After the first few rooms have been completed, the crew will most likely have the sequence memorized.

Activity 32: Develop Prefabrication Drawings for Field Use

- For systems that are partially or completely prefabricated in a prefabrication shop, develop drawings that show how the prefabricated parts should be assembled and installed in the field. Drawings that resemble standard product assembly instructions are particularly useful for field assembly and installation. These drawings should show each piece (or prefabricated piece) and how it connects to every other piece to form a complete system. Step-by-step assembly and installation instructions are especially helpful.

- For systems prefabricated in a shop, ensure the prefabricated parts are clearly labeled and that these labels correspond to an assembly and installation scheme identified on the drawings. Tag each prefabricated piece and match the tag to the step-by-step assembly and installation instructions. The idea is to make the installation of the prefabricated parts as simple as possible to decrease the amount of time it takes to install the system.
- If on-site prefabrication is scheduled, the drawings should identify the step-by-step process of how each piece is assembled in the prefabricated system.
- If off-site prefabrication is scheduled, separate drawings should also be created to identify how the prefabricated parts should be installed. Step-by-step installation instructions should be developed in conjunction with the prefabrication instructions.
- Distribute the prefabrication drawings to the foremen or field crews who will perform the work.

Schedule Development

Schedule development involves converting the sequencing and installation plan into a set of discrete work processes that can be mapped on a timeline. While a separate bar chart schedule is highly recommended, many specialty contractors have successfully added their own line items to the general contractor's bar chart schedule and essentially integrated their own schedule into the general contractor's schedule. A project schedule—whether independent or integrated—is essential for organizing the work and monitoring progress. The schedule development process consists of five activities.

Activity 33: Review General Contractor's Schedule and Timeline

- Complete the general contractor's schedule review items checklist shown in Figure 1.19.
- Initiate or attend a customer schedule review meeting to identify any special requirements and clarify any questions. Although schedule review meetings are not common in the construction industry, they are an effective way to clarify any special requirements, such as the timing of delivery of owner-furnished materials or the placement of heavy items on the roof (which would cause a temporary disruption in the work).

Completed	Item	Annotations
☐	Identify the overall project start date	
☐	Identify your start date	
☐	Identify the overall project completion date	
☐	Identify your completion date	
☐	List any interim milestones	
☐	Is the project divided into phases?	☐ Yes ☐ No
☐	Identify the start and completion dates for each phase	
☐	Does the contract include a liquidated damages clause?	☐ Yes ☐ No
☐	Does the contract include an incentive for early completion?	☐ Yes ☐ No
☐	Can your work be completed in the time frame identified in the contract documents?	☐ Yes ☐ No
☐	If the schedule must be compressed, create a plan for compressing the schedule and completing the work by the contractual completion date	
☐	Identify unusual scheduling requirements (e.g., night work, second shift work, after-school work hours, escort required)	
☐	Will the owner furnish any items, such as materials or equipment?	☐ Yes ☐ No
☐	Identify the "deliver no later than" dates that owner-furnished items must be delivered to the jobsite	
☐	Will any portion of the work be installed by the owner's own workforce or a separate contract?	☐ Yes ☐ No
☐	Identify the "install no later than" dates that owner-installed items must be completed	
☐	Identify techniques that will speed up the completion of the work	
☐	Review the sequencing/installation plan to determine how your work fits into the overall project schedule	
☐	Identify work that has already been completed	
☐	Identify work in progress and percent complete	
☐	Identify other contractors' activities that must be completed before your portion of the work can begin	
☐	Are you aware of any pending changes to the overall project or your work?	☐ Yes ☐ No
☐	If yes, describe the pending changes	
☐	Do you anticipate crowded site conditions? If yes, consider scheduling some of the work during a second shift to improve productivity	☐ Yes ☐ No

Figure 1.19 General contractor's schedule review items checklist

Activity 34: Identify Work That Impacts Your Activities

- While reviewing the overall construction schedule, annotate work that must be coordinated with other trades. For example, although the heating system is furnished and installed by the mechanical contractor, the heating unit must be supplied with a power source, which will be connected by the electrical contractor. The electrical contractor should identify all of the equipment in the facility that will require an electrical connection, which they must provide.

- While reviewing the overall construction schedule, annotate potential conflicts that will require coordination. For example, conflicts often occur between the mechanical, plumbing, and electrical piping, which must all fit in wall or ceiling space. Identify potential conflicts among systems that can be addressed before installation begins.
- Develop and submit RFIs to resolve open questions or conflicts among systems.
- Request an initial (and a recurring) coordination meeting to identify and resolve schedule questions and conflicts before installation begins. Many general contractors hold weekly coordination meetings to discuss upcoming work and to identify, discuss, and resolve conflicts among the various trades working on the jobsite. These meetings are usually effective at resolving issues expeditiously. However, if the general contractor has not arranged regular coordination meetings, request that the meetings be held, including an initial meeting that might require more time to identify all potential conflicts and issues plus possible solutions.

Activity 35: Review the Work Sequence and Long-Lead-Time Material/Equipment Delivery Dates

- Develop a rough draft of your schedule from the sequence and installation plan. The following rules of thumb are useful for developing the draft schedule:
  - Seek scheduling input from the field superintendent and foremen. They often understand how the work will be accomplished and how the installation method corresponds to the crew size and duration.
  - Develop the first draft schedule using actual durations to determine whether your work can be completed by the contractual completion date. If the schedule must be compressed, build the compression into the second draft schedule.
  - Consider developing a resource-loaded schedule so that the expenditures from the schedule can be compared to the expenditures from the cost report.

- Compare the estimated work hours for an activity (or work process) to the scheduled duration of that activity (or work process). Then either adjust the duration based on estimated work hours and expected crew size or adjust the crew size based on the estimated work hours and allowable duration.
- Be sure the schedule takes into account the time of year when the work item will be performed and the possible weather that can be expected. Excessive heat, cold, humidity, rain, or wind can significantly impact the progress of the work.
- Understand when major materials and equipment items will be delivered to the jobsite. The schedule may need to be adjusted to accommodate the delivery of those items.
- Be sure the schedule clearly documents the date when owner-furnished materials and equipment must be received. Also, make sure the owner knows when these items are required, and seek their acknowledgment in writing.
- Review the schedule of other specialty contractors to identify work that will impact your schedule.
- Your schedule should incorporate the work of any subcontractors. Review and incorporate your subcontractors' work into your own schedule and be sure to provide updates if the schedule changes.
- If crowded conditions or overmanning is expected on a jobsite (due to a small site or an acceleration of the work), consider scheduling some of your work during a second shift to improve work flow and increase productivity.
- Once your schedule has been reviewed and approved by the general contractor, diligently document delays and changes to the sequence caused by others. Thorough documentation can improve your chances of receiving a time extension and financial compensation.
- When developing work activities or elements for the schedule, review the budget, cost codes, and installation sequence plan. Create work elements that comprise no more than five percent of the scope of work so that inaccuracy in tracking progress will not have a significant impact on the percent complete.
- Consider creating a computerized bar chart schedule. Tracking and control will be greatly simplified if the updates can be made in a computer system.
- Consider dividing the schedule into areas or systems, such as first floor, second floor, and so on. This will simplify tracking and control, especially on larger projects.

- If some of the work is repetitive, consider creating these activities in your computer scheduling software and simply cutting and pasting as many times as the work occurs. The dates and durations can then be modified for each area of work.
- Verify the ordering and delivery dates of long-lead-time materials and equipment.
- Perform three-way coordination between sequencing/installation plan, material/equipment delivery plan, and the draft schedule. Set aside adequate time to coordinate the deliveries, installation sequence, and schedule. Annotate discrepancies that must be resolved when developing the final draft schedule.
- Modify the draft schedule and sequencing/installation plan to accommodate long-lead-time deliveries. The modifications should be coordinated so that the sequencing/installation plan matches the schedule.

Activity 36: Coordinate Your Schedule with the Schedule of the General Contractor

- After coordinating the sequencing/installation plan, material/equipment delivery plan, and the draft schedule, coordinate the draft schedule with the general contractor's schedule and adjust as necessary. A final coordination exercise between your schedule and the general contractor's schedule should identify any minor adjustments that might need to be made before producing your final schedule.
- Review your final schedule with the customer, general contractor, other trade contractors, and suppliers to resolve any final conflicts.
- Seek approval of the schedule from the customer and general contractor. Meeting minutes or a sign-off sheet that documents the customer's and general contractor's review and approval of your schedule is recommended. Simply sending the final schedule to the general contractor may be inadequate as acknowledgment of approval. Focus, instead, on a written signature.
- Seek integration of your schedule into the general contractor's overall project schedule. General contractors often show little detail about subcontracted trade work. Ask the general contractor to add all of the line items from your schedule into the overall project schedule, and provide the general contractor with a digital file, if available, to simplify the task.

Activity 37: Create a Bar Chart Schedule

- In addition to the integration of your schedule into the overall schedule, format your final schedule into an independent bar chart for tracking and control.
- Select the type of bar chart schedule to develop and track. The three most common bar chart schedules include: (1) simple bar chart schedule (which does not calculate the critical path); (2) critical path method (CPM) schedule; and (3) resource-loaded schedule that associates materials, equipment, labor, and cost with each schedule line item. Select the scheduling method that is most appropriate for the complexity of the project and skill of the scheduler.
- Save the original approved schedule as the baseline so that progress can be tracked and delays can be documented. The baseline schedule should be the final schedule that was approved by, and distributed to, the general contractor. It should never be changed. Instead, schedule updates that reflect the progress of the work and any accelerations, delays, or other changes to the work sequence or duration should be made routinely (weekly or monthly). An accurate baseline and final schedule can be used to support time extensions, change orders, and disputes.
- Distribute your bar chart schedule to the customer, general contractor, various subcontractors, and suppliers. The reason for distributing your final approved schedule is to provide official notification to all team members regarding how you intend to proceed with the work. It also provides suppliers with a tool for determining delivery dates; as a result, all updates to the schedule should also be distributed to team members.

Tracking and Control

Tracking and control involves selecting the proper control tools and setting up the company computerized project management system to track progress. It also involves developing a labor and materials tracking report and creating other essential reports in the computer database. The tracking and control process consists of two activities.

Activity 38: Customize the Computerized Tracking and Control System (e.g., Database/Schedule) for the Current Project

- Review the following tracking and control best practices:
  - Remember that tracking and control involves a time-cost trade-off. The more detailed the tracking and control system is, the

more time must be spent on entering data, and the greater the overhead costs will be. You must balance detail with simplicity.

- Be sure that you take into account the financial information needs of banks and bonding companies when setting up your control system.
- If your company keeps track of historic costs, be sure the tracking and control system captures the correct data in the proper format so that it can be added to the historic cost database.
- Match budget/cost control line items with field work items so that labor hours can be easily reported and tracked.
- Review the daily labor reporting procedures with the field supervisor and foremen. Be sure they understand how to code time cards so that the work performed can be directly associated with line items in the budget and tracking reports.
- Deviations between actual and estimated costs and work hours should be discussed with the field supervisor each week to identify and correct problems immediately.
- Seek feedback from the crew members about the causes of poor productivity and cost/labor hour overruns.
- Work together with crews to improve productivity. Often, management actions or inactions contribute to productivity outcomes. For example, poor instructions can hinder productive work, while speedy responses to questions can help increase productivity.
- If the crew members complete the work in fewer hours than estimated, provide a reward to thank them for their hard work, such as free lunch or special recognition.
- Try using incentives regularly to improve productivity.
- Make sure your data entries are as accurate as possible. Remember that cost reports should provide the project manager with a realistic financial overview of the project.
- A negative variance between estimated and actual cost and work hours should be addressed through corrective action. One of the main benefits of monthly cost reporting is to identify problems early enough to take corrective action.
- Complete the tracking and control tools checklist shown in Figure 1.20.
- Select computerized tracking and control tools for the project. The tracking and control checklist identifies several control tools that can be used to monitor project progress. Use the best practices and the

Completed	Completion date	Item
☐		Identify the goals of your tracking and control system. Do you want to: a. Monitor profitability? b. Identify variations in costs and work hours? c. Track productivity of the workforce? d. Contribute to your historical costs database? e. Document costs that are beyond the initial scope of work? f. Track changes in cost and work hours? g. Create contractually mandated cost reports for the customer? h. Evaluate the effectiveness of your management team? i. Conduct risk analyses on future projects of a similar type?
☐		Create a project schedule and format it so that you can update the progress of all schedule line items (see Activity 37). Match schedule line items to budget line items to facilitate effective tracking.
☐		As an alternative, create a resource-loaded schedule that will permit you to track cost and labor hours as you track schedule progress
☐		Create a manpower loading chart that identifies your crew size and composition for each week of the project
☐		Customize the computerized project management program so that you can use it as a tool to track RFIs, submittals, purchase orders, deliveries, and change orders
☐		Create a progress report to track variances in costs and labor hours. Your progress report should track budgeted line items.
☐		Create a labor productivity report to compare estimated productivity to actual productivity and identify solutions to problems. Daily timesheets should be coded to match budget line items.

Figure 1.20 Tracking and control tools checklist

checklist to determine which tools are appropriate for the current project.

- Modify the tracking and control tools for the current project. Use the generic computer tracking and control tools to input data about the current project. Then, format the control tools so that they can be used to track and control progress on the project.

Activity 39: Develop Labor and Materials Tracking Report

- Review the cost estimate and associated budget and the cost code scheme.

Day	Work description	Cost code	Hours		
			Regular	Overtime	Double time
Mon					
		Total hours			

Day	Work description	Cost code	Hours		
			Regular	Overtime	Double time
Tue					
		Total hours			

Day	Work description	Cost code	Hours		
			Regular	Overtime	Double time
Wed					
		Total hours			

Day	Work description	Cost code	Hours		
			Regular	Overtime	Double time
Thu					
		Total hours			

Day	Work description	Cost code	Hours		
			Regular	Overtime	Double time
Fri					
		Total hours			

Day	Work description	Cost code	Hours		
			Regular	Overtime	Double time
Sat					
		Total hours			

Figure 1.21 Daily labor time report

- Match the cost estimate/budget work items to crew assignments so that labor and materials can be tracked easily. Using the estimate and budget as a guide, verify that labor and materials can be easily reported daily/weekly for each line item in the budget. If the reporting process is too complicated or detailed, consider revising the budget to simplify the tracking and control process. Figure 1.21 provides a sample time card/labor reporting form that links daily labor to specific cost codes in the budget, which can then be tracked using a progress report, labor report, or productivity report.
- Select the reports to be created and used for monitoring progress. The two most common reports include a progress report as shown in Figure 1.17 and a productivity report as shown in Figure 1.22.
- Develop the selected reports by inputting project data into the reporting system (database or accounting system). Once the tracking and control reports have been selected, project-specific data should be entered into the database to populate the reports. These reports can then be used to update progress and productivity on a daily, weekly, or monthly basis.

Construction Execution Kickoff Meeting

The final step to pre-construction planning is to hold a construction execution kickoff meeting that calls together all internal team members to review communication processes, administrative procedures, reporting requirements, and project budgets and schedules immediately prior to executing the work. The purpose of the kickoff meeting is to ensure all pre-construction planning tasks have been completed and to prepare the team to execute the project. The construction execution kickoff meeting should begin by selecting a date and location for the meeting, inviting those team members who attended the planning kickoff meeting, and distributing the agenda. Figure 1.23 shows a sample agenda. The construction execution kickoff meeting consists of seven activities.

Activity 40: Review Meeting Schedule

- Review the schedule of weekly internal meetings associated with the project. Meetings that should be held include (as a minimum) a weekly on-site progress meeting (including a walk-around) and an in-office staff meeting that covers the progress on all projects being performed by the company. Other meetings that might be necessary are specialty subcontractor coordination meetings (if not conducted by the general contractor), look-ahead planning meetings, and safety meetings.

CSI division	Description	Units	Estimated					Actual				
			Qty of units	Hours	Hours/ Units	Total cost		Qty of Units	Hours	Hours/ Units	Total cost

Figure 1.22 Labor productivity report

- Review the schedule of weekly project meetings conducted by the customer or general contractor. Typically, two types of meetings are held: progress meetings and coordination meetings. The purpose of progress meetings is to update the customer/general contractor on the progress of work and to discuss and resolve open RFIs, pending changes, submittals, and other issues. The purpose of coordination meetings is to assemble all trades working on the jobsite to discuss upcoming work, potential conflicts, and to identify solutions to the

1. Project overview, including:
   a. Project name
   b. Location
   c. Type of work
   d. Contract cost
   e. General scope of work

2. Introduce internal team members and any changes in team members since planning meeting, including:
   a. Project manager
   b. Field supervisor
   c. Foremen
   d. Estimator
   e. Accounting representative
   f. Purchasing agent
   g. Director of operations
   h. Other internal team members

3. Identify external team members and any changes in team members since planning meeting, including:
   a. Owner/Customer
   b. Architect/Engineer
   c. General contractor/Construction manager
   d. Other specialty subcontractors
   e. Vendors/Suppliers

4. Review the general scope of work
   a. Provide an overview of the entire project scope of work and your specific scope
   b. Review major work performed by others

5. Review the meetings schedule, including:
   a. Internal office and jobsite progress meetings
   b. Jobsite safety and coordination meetings
   c. Project progress and coordination meetings
   d. Other meetings

6. Review the RFI process, including:
   a. Standard company RFI procedure
   b. Contract language regarding RFIs and the RFI process
   c. Current outstanding RFIs

Figure 1.23 Construction execution kickoff meeting agenda

7. Review the change order process, including:
   a. Contract language regarding changes and the change order process
   b. Standard company change order and field change processes
   c. Documentation of changes, delays, disruptions, and disputed work

8. Review the submittal process, including:
   a. Contract language regarding submittals and the submittal process
   b. Standard company submittal processes
   c. Following up on late submissions and approvals

9. Review the billing process, including:
   a. Contract language regarding billing and the payment process
   b. Standard company billing processes
   c. Subcontractor/Supplier/Vendor invoicing and payment
   d. Following up on late payments or open accounts receivable

10. Review the tracking and control process, including:
    a. Labor reporting
    b. Progress reporting and percent complete
    c. The progress update timeline
    d. Standard company progress update procedures

11. Review the schedule and milestones, including:
    a. Customer-furnished schedule
    b. Bar chart schedule
    c. Work by others that will impact your work
    d. Important material and equipment delivery dates
    e. Schedule updating to reflect percent complete

12. Review site logistics and material storage and staging, including:
    a. Site access
    b. Parking
    c. Material delivery procedures
    d. Material storage locations
    e. Trailer locations
    f. Site cleanup requirements
    g. Temporary power and lighting requirements

13. Review special safety issues

14. Review other items specific to this project

Figure 1.23 *Continued*

conflicts. Additional meetings that might be required include look-ahead scheduling meetings and safety meetings.
- Identify the internal team members who will be responsible for attending the internal and external meetings.

Activity 41: Review Request for Information Process
- Review the contract for a customer-mandated RFI process. If the contract identifies a customer-mandated RFI process, you will be bound

to follow the process. If no RFI clause exists, you should convey your company standard procedures to the general contractor.

- Review the company standard procedure for developing, processing, tracking, and closing out an RFI.
- Modify the company standard RFI procedure, as necessary, to conform to the contractually mandated process. Distribute the procedure so that all employees who might submit an RFI understand the submission process.

Activity 42: Review Change Order Process and Field Change Management Process

- Review the contract for a customer-mandated change order process and field change request process. The contract should be carefully reviewed for clauses that identify when compensation for changes will be awarded and the type of compensation permitted (time, cost, or both). Many clauses have a notice requirement, and some clauses only permit time extension without financial compensation. Figure 1.7 provides a checklist to guide team members through the contract review process.
- Review the company standard change order procedure and field change request/management process.
- Discuss the process for documenting changes, delays, and disruptions in the work flow and sequence. Using the company change order process as a guide, the team members should discuss how to accurately document impacts to the installation process in order to approach the customer or general contractor for compensation. Keeping accurate records will be essential to the recovery process and should be a top priority for the project manager, field supervisor, and key foremen.
- Discuss the process for tracking and following up on change order requests and payment for field change directives. Responsibility for managing change orders and following up on their approval is typically assigned to the project manager. Responsibility should be clearly assigned during the meeting, and the tracking process should be reviewed.

Activity 43: Review Submittal Processing Procedure

- Review the contract documents for customer-mandated submittal processing procedure. The specifications may provide a submittal log (either blank or populated with the required submittals) and a set of submission procedures that must be followed. Some general contracts may require complete books with all submittals be processed as a whole, while others may allow submittals to be processed individually. If no submittal process is identified in the contract documents, you should convey your company standard procedures to the general contractor.

- Review the company standard procedure for developing, processing, tracking, and receiving approval for project submittals. Activity 17 discussed the setup of a submittal processing and tracking procedure, and Activity 24 discussed the process for requesting submittals from suppliers.
- Modify the company standard submittal procedure, as necessary, to conform to the contractually mandated process.

Activity 44: Review Billing and Invoicing Procedures

- Review the general contractor's process and timeline for invoicing, lien waivers, and payment. Identify the date each month that invoices must be submitted to the general contractor in order to be processed during the regular pay cycle. Also make sure to review the terms of payment, such as whether the invoice will be paid in 30, 60, or 90 days.
- Review the internal billing cycle established by the accounting department. Activity 28 described the budget development process and Activity 29 described the development of the schedule of values. The budget, schedule of values, and company billing process should be reviewed and compared to the general contractor's invoicing cycle to make sure payments can be processed during the general contractor's regular invoicing cycle.
- If necessary, adjust the internal billing cycle to conform to the general contractor's invoicing cycle. If the company billing cycle must be modified to conform to the general contractor's invoicing cycle, be sure to document and distribute the changes to all internal team members who will be involved in the billing process. Also notify all subcontractors, suppliers, and vendors of the billing and payment process for the project.
- Discuss the process for following up with the general contractor on late unpaid invoices. Typically, the accounting department will request that the project manager assume responsibility for following up on late payment or unpaid invoices. The follow-up process should be discussed, and responsibility should be clearly assigned.

Activity 45: Review Project and Field Reporting and Tracking Procedures

- Discuss the process for reporting labor hours and submitting time cards to the accounting office. In a typical process, the crew foreman will record the hours worked by each of the crew members. The foreman will submit the weekly hours to the field supervisor, who will forward them to the project manager. The project manager will review and approve the time cards and forward them to the accounting department for entry into the accounting system. The accounting department will enter the hours, and the hours can then be tracked by the progress and productivity reports. The reporting process should be clarified in the meeting.

- Discuss the process for reporting progress and percent complete. Typically, the project manager and field supervisor will establish the percent complete together. The process requires the project manager to understand how much work was supposed to be completed on a particular date and then walk around with the field supervisor to determine how much work has actually been completed. Since each work element should account for no more than 5 percent of the total work, minor inaccuracies should not significantly skew the total percent complete.

- Discuss the process for tracking material and equipment costs and purchase orders. Typically, the accounting department will enter payment information into the accounting database when purchase orders or invoices are paid. It is important that these payments are correctly allocated to the proper cost code so that the progress report accurately reflects how much has been spent for each line item in the budget. The purchase order payment entry process should be clarified in the meeting.

- Review and discuss the tracking tools that will be used to monitor progress. Several possible tools that can be used to track progress include the progress report, productivity report, schedule, and manpower loading chart. The company may have additional tools that are used routinely. The project manager should identify the tools that will be used and discuss the process for updating each control tool.

- Review the monthly update process. The company standard process should be reviewed during the meeting.

Activity 46: Review Your Schedule and the Schedule of the General Contractor

- Review your bar chart schedule. Activity 37 discussed the process for creating your schedule. The final schedule should be reviewed so that all internal team members are familiar with the timeline and sequence of activities.

- Review the general contractor's schedule and/or timeline. Activity 33 discussed the process for reviewing the general contractor's schedule. A brief overview of the general contractor's overall project schedule is sufficient so that all team members understand the general scope of the project and the completion timeline.

- Discuss the process for updating the schedule. Typically, the schedule is updated by the project manager. At the end of each week, the project manager and field supervisor should walk around the jobsite to compare the actual work completed to the work that was expected to be completed. Completion can be determined by quantities of units installed or approximate percent of a system installed. The schedule can then be updated based on percent complete of the schedule line items or number of units installed (if the schedule is resource-loaded or entered in sufficient detail).

PLANNING ASSESSMENT PROCESS

Our research discovered that projects that implemented a planning process similar to the model process tended to perform more successfully. Furthermore, there appeared to be 16 activities that had an especially strong influence on performance—and many of these activities were often overlooked during the planning process. The purpose of this section is to present a scorecard that can be used to evaluate the effectiveness of the planning that was completed on a project that is about to be executed. This section begins by identifying the 16 activities that had an especially strong influence on performance in order to explain why these activities are weighted more heavily on the planning scorecard. Then, the planning effectiveness scorecard is introduced, which includes all of the activities from the model planning process, and instructions are provided on how to score the effectiveness of planning on a new project. Finally, a section on score analysis presents some rules of thumb for projects that have various characteristics, such as large projects versus small projects, complex projects versus simple projects, and so on.

Important and Influential Planning Activities

It is essential to note that each of the 46 pre-construction planning activities in the model pre-construction planning process is an important part of the planning process. It is also worth noting that not every activity will require a great deal of time to complete—indeed, many of the activities can be completed in a matter of a few minutes, particularly on smaller projects. The project manager will need to exercise skill and judgment when determining how much time to devote to each activity.

Many of the activities are clearly critical to an effective planning process. Such obvious activities as reviewing the plans and specifications, administratively setting up the project, and issuing purchase orders can directly impact the smooth execution of a project. Overall, 16 influential activities were identified through careful research. It is critical that the project manager understand that all of the 46 pre-construction planning activities are important to project success. However, the following 16 activities have a particularly strong influence on the outcome of a project:

- Team selection and turnover
    - Activity 2: Hold turnover meeting between estimator and project manager
    - Activity 3: Hold separate turnover meeting between project manager and field supervisor
- Scope and contract review
    - Activity 7: Field supervisor reviews plans, specifications, and schedule
    - Activity 10: Compare estimated (bid) work activities and materials to planned performance
    - Activity 11: Identify value engineering and prefabrication opportunities and how to simplify the work
- Administrative setup
    - Activity 16: Initiate a request for information tracking and processing system
    - Activity 17: Initiate a submittal tracking and processing system
- Buyout process
    - Activity 23: Request submittals, cut sheets, and shop drawings
- Material handling plan
    - Activity 25: Develop material delivery and handling plan
    - Activity 26: Develop material storage and staging plan
- Layout and sequencing plan
    - Activity 30: Develop installation sequence and layout drawings
    - Activity 31: Develop field instructions
    - Activity 32: Develop prefabrication drawings for field use
- Schedule development

- Activity 36: Coordinate your schedule with the schedule of the general contractor
- Construction execution kickoff meeting
  - Activity 43: Review submittal processing procedure
  - Activity 44: Review billing and invoicing procedures

When performing pre-construction planning, extra care should be taken to include these activities in the planning process. Although it is recommended that all of the 46 activities be completed, under extraordinary circumstances when the planning process must be abbreviated, these 16 activities should be among those that get completed.

Planning Effectiveness

The planning assessment process provides a simple method for evaluating whether each of the 46 pre-construction planning activities has been completed, and whether an activity was completed before or after a project was executed. Under ideal circumstances, all of the planning activities will be completed prior to executing the work. However, ideal circumstances are rare in the construction industry, and often some of the planning activities must be completed after the project has been executed.

The planning effectiveness scorecard was developed as a tool to evaluate how closely a project's actual planning process matches the model pre-construction planning process. The scorecard, shown in Figure 1.24, lists each of the 46 model pre-construction planning activities (column C). Space is provided in column D for contractors to identify whether or not the activity was performed and to assign a performance score. The performance score is assigned based on whether an activity was performed or not performed, and whether it was completed before or after execution using the following scale:

- Assign two points if the activity was completed before executing the work
- Assign one point if the activity was completed after executing the work
- Assign no points if the activity was not completed

Using this scale, a higher summed performance score indicates that more of the pre-construction planning activities were completed before executing the work, which correlates strongly to better project performance. Column E represents the weight assigned to each activity. As mentioned earlier, 16 of the activities have a stronger influence on performance, and, as a result, these activities are assigned larger weights. To calculate a final score (column F) for each activity, column D

Category (A)	Activity number (B)	Activity (C)	Performance score (D)	Weight (E)	Final score (F) = (D)×(E)
Team selection and turnover	1	Finalize selection of project manager, field supervisor, and other key team members		1.00	
	2	Hold turnover meeting between estimator and project manager		2.00	
	3	Hold separate turnover meeting between project manager and field supervisor		4.00	
	4	Hold pre-job (planning) kickoff meeting with internal team members to assign responsibilities		1.00	
Scope and contract review	5	Review contract for unfavorable or high-risk clauses		1.00	
	6	Project manager reviews plans, specifications, and schedule		1.00	
	7	Field supervisor reviews plans, specifications, and schedule		3.50	
	8	Create a list of issues that need to be resolved and begin the RFI process		1.00	
	9	Conduct site visit		1.00	
	10	Compare estimated (bid) work activities and materials to planned performance		3.00	
	11	Identify VE and prefabrication opportunities and how to simplify the work		2.00	
	12	Prepare construction takeoff		1.00	
Administrative setup	13	Set up project files and create contact list		1.00	
	14	Set up computerized tracking and control system (forms, database, schedule, tracking)		1.00	
	15	Initiate a change management system		1.00	
	16	Initiate an RFI tracking and processing system		2.00	
	17	Initiate a submittal tracking and processing system		3.50	
	18	Develop a labor requirements/expectations letter		1.00	
Buyout process	19	Request and/or review subcontractor/supplier/vendor prices and qualifications		1.00	
	20	Negotiate pricing and contract conditions and select subcontractors/suppliers/vendors		1.00	
	21	Develop and issue purchase orders and contracts for materials and equipment		1.00	
	22	Order long-lead-time materials and equipment		1.00	
	23	Request submittals, cut sheets, and shop drawings		2.00	
	24	Develop and process log and book of submittals, cut sheets, and shop drawings		1.00	

Figure 1.24 Planning effectiveness scorecard

(continues)

Category (A)	Activity number (B)	Activity (C)	Performance score (D)	Weight (E)	Final score (F) = (D)×(E)
Material handling plan	25	Develop material delivery and handling plan		2.50	
	26	Develop material storage and staging plan		2.00	
Budget preparation	27	Develop, review, or expand cost code scheme		1.00	
	28	Develop budget by breaking down labor, material, overhead, and profit costs		1.00	
	29	Develop schedule of values		1.00	
Layout and sequencing plan	30	Develop installation sequence and layout drawings		2.50	
	31	Develop field instructions		3.00	
	32	Develop prefabrication drawings for field use		2.50	
Schedule development	33	Review general contractor's schedule and timeline		1.00	
	34	Identify work that impacts your activities		1.00	
	35	Review the work sequence and long-lead-time material/equipment delivery dates		1.00	
	36	Coordinate your schedule with the schedule of the general contractor		2.00	
	37	Create a bar chart schedule		1.00	
Tracking and control	38	Customize the computerized tracking and control system (e.g., database/schedule) for the current project		1.00	
	39	Develop labor and materials tracking report		1.00	
	40	Review meeting schedule		1.00	
	41	Review RFI process		1.00	
Construction execution kickoff meeting	42	Review change order process and field change management process		1.00	
	43	Review submittal processing procedure		3.50	
	44	Review billing and invoicing procedures		3.00	
	45	Review project and field reporting and tracking procedures		1.00	
	46	Review your schedule and the schedule of the general contractor		1.00	
		Planning effectiveness score (sum of column F)			

Figure 1.24 *Continued*

(performance score) is multiplied by column E (weight). The planning effectiveness score is the sum of all of the total scores.

Score Analysis

Projects that were investigated as part of the research had various characteristics. In particular, four characteristics were analyzed, including (1) project size, (2) initial uncertainty, (3) bid accuracy, and (4) type of construction.

Project size. This variable is generally determined by three values: (1) contract cost at award; (2) original estimated total work hours; and (3) estimated peak number of craftsmen. Furthermore, estimated project duration is also strongly related to a project's size. The trend seems to indicate that larger projects will require a higher planning effectiveness score to have a greater chance for successful performance.

Initial uncertainty of the project. Initial uncertainty of the project is generally determined by three values: (1) perceived level of uncertainty (high, medium, or low); (2) percentage of the total design completed at bid; and (3) perceived level of complexity (high, medium, or low). The trend indicates that projects with high levels of initial uncertainty will require a higher planning effectiveness score to have a greater chance for successful performance.

Bid accuracy. This variable is generally determined by two concepts: a perceived accurate cost estimate and perceived accurate estimated work hours. The trend indicates that projects with an inaccurate bid will require a higher planning effectiveness score to have a greater chance for successful performance.

Type of construction. This variable is generally determined by classifying projects as (1) commercial, (2) industrial, (3) institutional, and (4) other. The trend indicates that industrial and institutional projects require a higher planning effectiveness score to have a greater chance for successful performance.

CONCLUSIONS

Our research provided evidence to support the hypothesis that the projects that experience an appropriate planning effort also experience more successful outcomes. The phrase "appropriate planning effort" was ultimately defined as planning that takes into account the inherent characteristics of a project, where inherent characteristics include (1) project size, in terms of estimated cost, estimated work hours, estimated number of workers, and duration, (2) initial uncertainty of the project, in terms of perceived uncertainty of the work, completeness of the design, and the perceived complexity, (3) bid accuracy, in terms of the accuracy of the cost estimate and the accuracy of the estimated work hours, (4) existing relationships, in terms of whether the subcontractor has worked with the owner or general contractor before, and (5) type of construction and award, in terms of type of work and whether the subcontractor was the low bidder on projects.

It was ultimately discovered that the projects that performed more effective planning (or performed planning activities similar to the model planning process) and had lower uncertainty and bid inaccuracy tended to perform more successfully. Indeed, various combinations of project characteristics and effective planning resulted in various chances of achieving a successful outcome. Therefore, both planning effectiveness and inherent characteristics should be evaluated together to identify the likelihood of successful performance.

Evidence was also provided to support the hypothesis that there are significant pre-construction planning activities that distinguish a successful project from a less-than-successful project. Specifically, 16 activities that significantly affected the performance of a project were identified.

Our research resulted in the creation of a model pre-construction planning process that was based on outstanding processes used on successful projects. The model process is a standardized process that contractors can implement to formalize their planning practices and improve the effectiveness of their planning. Furthermore, our research developed a method to assess planning effectiveness by comparing actual planning to the model process and filling out a scorecard. The scorecard can be used by contractors to identify specific planning activities that need additional effort prior to project execution.

WEB ADDED VALUE
Planning.doc

This file provides checklists for each one of the 10 categories that make up the model pre-construction planning process. In addition, standard procedures for change orders, RFIs, submittals, transmittals, billing, and progress updates are also provided. Finally, a sample requirements and expectations letter and a sample letter of intent complete the package. These checklists, generic procedures, and sample letters are intended to provide a starting point for a contracting firm to develop its own documents to implement the model pre-construction planning process presented in this chapter.

Web
Added
Value™

This book has free material available for download from the
Web Added Value™ resource center at *www.jrosspub.com*

2

EARLY WARNING SIGNS
OF PROJECT DISTRESS

Dr. H. Randolph Thomas, *The Pennsylvania State University*

INTRODUCTION

There is some belief in the construction industry that project difficulties are the result of unforeseen events that are beyond the control of contractor management. Indeed, there is no shortage of excuses on a construction site. It is also believed that current project control systems are adequate to advise contractors of potential risks. Consequently, there is a limited understanding of the crisis exposure and risks (Loosemore and Teo 2000).

Senior managers are at times oblivious to the events happening on the job-site. These events can seriously affect the profit margin on a particular project. It is often said that the accountant delivers the dreaded news near the end of the project, and this is the first time it is known by home office managers and executives that the project is in trouble. Effective management must be known early in the project so that problems are rectified as they happen. Then the accountants will not have to deliver bad news later. Unfortunately, current schedule and cost control systems can be too cumbersome, detailed, late, and lacking the key information necessary to provide timely and easily decipherable clues or red flags of project distress.

The purpose of this chapter is to introduce a methodology for document-ing key events and information that will allow senior managers to quickly and efficiently observe situations unfolding that will potentially affect the specialty contractor's performance. A Project Distress Diary (PDD), is updated weekly and retained at the site, becomes the repository of the key happenings on the project. Review of the PDD will reveal when certain actions are appropriate to avoid schedule acceleration.

The information in the PDD is easy to assemble and is largely extracted from the daily diaries, project schedule, progress meeting minutes, and other readily accessible sources. It is not the purpose of the PDD to replace existing project control systems, but rather to supplement them.

The focus of the PDD is to provide early warning distress signs that will likely lead to schedule acceleration. It is important to examine the performance of other contractors who will ultimately affect your performance. Therefore, the emphasis of the PDD is on how well the project begins. Previous research and experience shows that if a project starts poorly, it is difficult to recover the lost schedule time (Thomas et al. 1985). The most likely deficiency is in contractor management. Management deficiencies do not go away, but instead manifest themselves in other ways. Therefore, the warning signs of management deficiencies evident early in the job are omens of things to come.

PROJECT DISTRESS DIARY

The components of the PDD are based on the Factor Resource Model shown in Figure 2.1. The figure shows that resources are needed to prosecute the work efficiently and in a timely manner. Therefore, one of the early warning signs is when resources are not available or are not being used as planned. Disruptions are inhibitors to the project and can include bad weather, congestion, changes, and out-of-sequence work. Where these inhibitors occur early in the project, their presence can be a harbinger of things to come. Therefore, the presence of selected disruptions is monitored. Progress is an important indicator. Therefore, it is necessary to track key activities whose timely completion is essential to the work of the specialty contractor. Thus, Figure 2.1 is an efficient framework from which to monitor early warning signs.

The Factor Resource Model is a simple but flexible model of factors affecting the work. It can be expanded as needed. For instance, there could be another set of boxes leading to work methods that are within the contractor's control, which could include supervision issues such as lack of foreman experience, inadequate planning and scheduling, craft turnover, or changes in supervision. These are controllable inhibitors, whereas the others are outside the control of the specialty contractor.

The PDD is maintained on-site by the senior site representative of the contractor. It should be updated weekly. The PDD should be available for home office

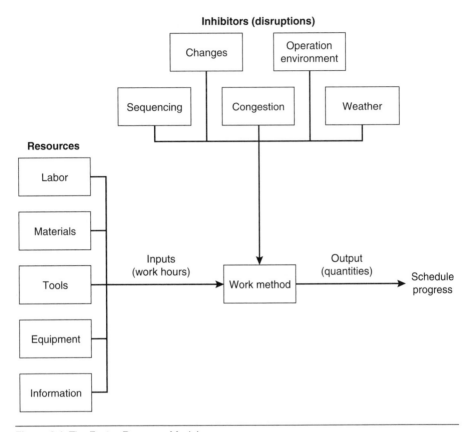

Figure 2.1 The Factor Resource Model

managers visiting the site. In a matter of minutes, it can be determined if there are indicators significant enough to warrant action to preclude future problems. The PDD does not isolate the root causes of the problem, but rather the symptoms. The root causes can be varied and numerous.

The PDD can also be used for other purposes. It can be applied to any type of project. Additionally, it can be incorporated in the training of foremen or as part of the training of project managers. It can also be used as part of a college course or co-op/intern training program where one would visit a job weekly to follow the sequence of the work and understand how the work is affected.

The PDD has been developed to be easy to maintain. It relies on readily available information from project schedules, the contractor's project control reports, and from the minutes of progress meetings. The design of the PDD is flexible and can be tailored to the uniqueness of each project. Therefore, where certain information is irrelevant or unavailable, it can be deleted.

The following sections illustrate the use of the PDD with two case studies applied to electrical contractors.

CASE STUDY 1

Case Study 1 is a three-story, steel frame laboratory/classroom building on the campus of a small university in central Pennsylvania. The schedule duration of the construction was 48 weeks. A time extension was unlikely, as the building was needed for the start of classes in the fall semester. The cost of the facility was $5.5 million. The type of contract for all subcontractors was a lump sum.

The PDD was updated weekly through week 32. Two evaluations are made. The first is at the end of week 25, which is near the end of the civil construction phase. This period is termed the documentation phase, and the contents of the PDD are illustrated using information from this phase. The second evaluation is at the end of week 32, and is termed the appraisal. The purpose of the appraisal is to determine if the warning signs made during the documentation phase have disappeared and are an aberration or if they were truly a warning sign of things to come.

Documentation Phase

The PDD is generally organized as in Figure 2.1. Other relevant information is also included. The PDD contains four sections as described in the following paragraphs.

Section 1: Project Participants and Key Events

In this section of the PDD, the principal participants and contract amounts are recorded. The rationale is that this information may not be known at the time the specialty contractor is awarded a contract, and prior involvement or knowledge of certain participants may yield clues as to how the project may be managed. If a specialty contractor has underbid his/her work, this information may also be relevant.

The key events in this section of the PDD involve recording the major events or conditions affecting the project. This is, in essence, a condensed version of a project diary. This chronology is not intended to replace the normal project diary, but rather to highlight relevant information that would alert others that something is not progressing as planned or expected. An important supplement to the chronology of key events is project photographs. The key events noted on the case study project are as follows:

- The footings and piers were complex. The plans were also difficult to interpret. Therefore, working drawings were prepared so that the work could proceed.

- There were late design modifications to the storm water system. This delayed the excavation of the foundation, partly because of permit requirements.
- Many underground, unforeseen utilities were found.
- Important submittals, especially structural steel, were late being approved.
- The housekeeping was sometimes marginal.
- The steel erection was done in two phases because of the size of the crane that was used by the subcontractor. Erection was done with one crane.

Photographs are particularly useful in assessing the project's progress.

Section 2: Progress

This section affords the opportunity to document the planned and actual progress on selected work that will most likely affect the specialty contractor. Many projects rely on the use of a CPM schedule. This information is often too detailed for a quick evaluation, especially as it may affect the specialty contractor's work. Instead, a more simplified approach is taken. The superintendent is asked to identify up to eight milestone activities that are necessary to be completed in a timely manner so as not to affect the specialty contractor's work. The focus is on the early work and covers both the civil and services/interior work. Examples of civil-related milestone activities include excavation, construction of footers and foundation, structural steel erection, and roof installation. The services/interior milestones may include plumbing and piping, rough-in electrical, insulation, drywall, and other activities. Since each project is unique, milestone activities are uniquely selected by the superintendent.

The planned period of performance for each activity can be readily obtained from the project schedule. The absence of a schedule or denial of access to this information is in itself a red flag of potential trouble. The actual period of performance can be obtained through personal observation or from the minutes of weekly or biweekly progress meetings. Other red flags include denial of the minutes or where the specialty contractor is barred from attending the meetings.

A milestone schedule from the case study project is shown in Figure 2.2. Notice from this figure that the milestone activities have been uniquely selected because of their potential effect on the specialty contractor's work that in this case happens to be an electrical contractor. The first three activities are civil/structural. As can be seen, the project as of week 25 was four weeks behind the planned schedule. The electrical rough-in has yet to start (it was supposed to start at the end of week 23), and is at least two weeks late.

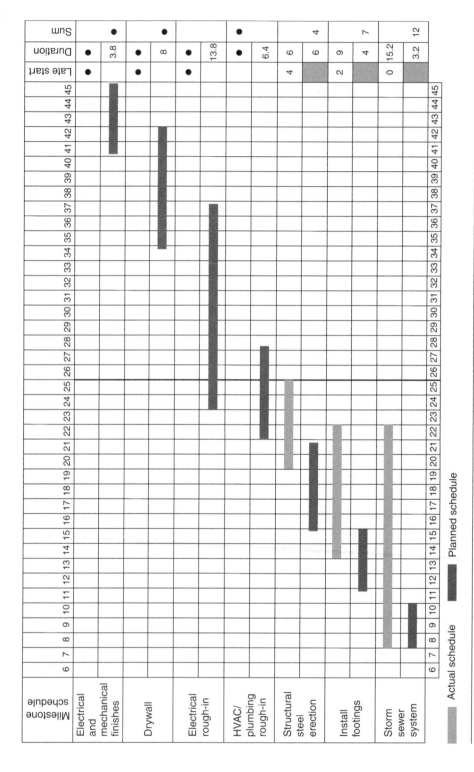

Figure 2.2 Milestone schedule as of week 25 for Case Study 1

Section 3: Resources

The resources documented in the PDD (see Figure 2.1) are:

- Labor
- Installed equipment
- Information
- Money

Deviations from the planned utilization of resources and delays are strong red flags that the project has been delayed or will be delayed in the future. Out-of-sequence work may be needed to avoid completion delays. Other, less disruptive steps may be possible, but this requires advance planning.

Labor. The actual percentage of weekly labor of the electrical contractor should follow the planned utilization rate. Where events occur that interrupt the schedule, the effects will appear as deviations from the plan. Figure 2.3 illustrates that the utilization of electrical labor is slow at the beginning of the work in response to the lack of footing and slab progress. This shortfall in utilization will likely need to be made up later through the use of overtime or overmanning. Unplanned labor utilization is a clear red flag of distress.

Installed Equipment. Another resource that can be easily tracked is the installed equipment. Contractor procured and owner procured equipment is included in the PDD, and only the key pieces of installed equipment are monitored. Contractor equipment can be the electrical equipment or key equipment of

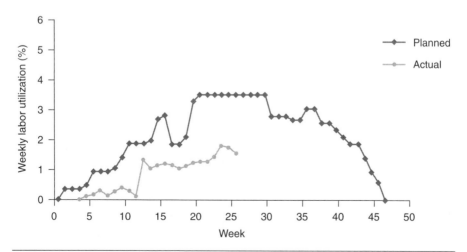

Figure 2.3 Hypothetical comparison of manpower consumption rates

other contractors. Figure 2.4 shows that only the elevator on the case study project was deemed significant enough to track. As seen, following approval, it took 14 days for the order to be placed. The elevator was delivered a month late.

Information. Submittals are one of the most important items of information to track. As with equipment, only those submittals deemed important should be monitored. Generally, this means keeping track of the submittals required of other contractors. The status of these submittals should be the subject of discussion at the weekly or biweekly progress meetings, and the status should be readily available from the minutes of that meeting. Figure 2.5 provides a summary of submittal information from the case study project. The list is longer than might normally be expected as there are 11 submittals related to the structural steel erection subcontractor. In reviewing Figure 2.5, there are a number of red flags that should be of concern to the electrical contractor. First, there is no required submission date for submittals. Thus, it is left for each subcontractor to produce submittals when they choose. The steel submittals are particularly troubling. These submittals are generally taking three to four weeks to approve. The erection activity has started 30 days late (see the last column in Figure 2.5). Several electrical submittals have yet to be approved. Overall, management of the submittal process seems to be less than satisfactory. The cause may be late submission, slow approvals, design errors, or other causes. Nevertheless, the general tardiness of approvals would suggest that schedule acceleration looms more likely as time passes.

Contractor procurred and other equipment

Item	Approve		Order		Deliver	
	Plan	Actual	Plan	Actual	Plan	Actual
Elevator		5/4/xx		5/18/xx		7/1/xx

Figure 2.4 Equipment procurement status for Case Study 1

Submittals ID	Original		Re-submission (3)	Final approval		Activity start date		Submittal lateness (5–4)	Activity lateness (6–7) days
	Schedule (1)	Actual (2)		Scheduled (4)	Actual (5)	Planned (6)	Actual (7)		
Storm detention vault, grates, piping shop drawings	NS	5/21/XX		NS	6/13/XX	NS	6/18/XX	NS	0
Rebar shop drawings 1–6	NS	5/31/XX		NS	6/15/XX	NS	9/25/XX	NS	(–)73
Structural steel shop drawings AB1, I	NS	5/22/XX		NS	6/15/XX	NS	9/10/XX	NS	(–)30
Structural steel shop drawings E1, E2, JS1, 2–16 and 26–33	NS	5/22/XX		NS	6/15/XX	NS	9/10/XX	NS	(–)30
Structural steel shop drawings E1, (rev #1) and E2 (rev #2)	NS	5/22/XX		NS		NS	9/10/XX	NS	(–)30
Structural steel shop drawings E1 (rev #2) and E2 (rev #2)	NS	5/23/XX		NS	6/22/XX	NS	9/10/XX	NS	(–)30
Structural steel shop drawings E2 (rev #3)	NS	5/23/XX		NS		NS	9/10/XX	NS	(–)30
Structural steel shop drawings E1 (rev #3)	NS	5/25/XX		NS		NS	9/10/XX	NS	(–)30
Structural steel shop drawings E2 (rev #4)	NS	5/25/XX		NS		NS	9/10/XX	NS	(–)30
Structural steel shop drawings E1 (rev #4)	NS	6/28/XX		NS		NS	9/10/XX	NS	(–)30
Structural steel shop drawings E2 (rev #5)	NS	6/25/XX		NS		NS	9/10/XX	NS	(–)30
Structural steel shop drawings E4, 55–58, 61–72, 34, 51–54	NS	5/25/XX		NS	6/22/XX	NS	9/10/XX	NS	(–)30
Steel joists shop drawings J1, J2, J3 and J4	NS	5/22/XX		NS	6/15/XX	NS	9/10/XX	NS	(–)30
Hydraulic elevator shop drawings	NS	6/22/XX		NS		NS		NS	
Underground HVAC distribution piping shop drawings	NS			NS		NS	11/21/XX	NS	(–)60
Laboratory air and vacuum equipment wiring diagrams	NS			NS		NS		NS	
Chemical waste system shop drawings	NS	6/04/XX		NS	6/13/XX	NS		NS	
Fans and ventilator shop drawings	NS	5/22/XX		NS	6/13/XX	NS		NS	
Pad mounted transformer shop drawings—square D	NS	5/16/XX		NS	6/04/XX	NS		NS	
Main distribution switchboard shop drawings—square D	NS	5/16/XX		NS		NS		NS	
Panel boards shop drawings	NS	5/16/XX		NS		NS		NS	
Packaged engine generator system shop drawings	NS	5/16/XX		NS		NS		NS	
Fire alarm and smoke detection system shop drawings	NS	6/20/XX		NS		NS		NS	
Telecom system cable tray/run away shop drawings	NS			NS		NS		NS	

NS—no schedule

Figure 2.5 Submittal schedule for Case Study 1

Money. Money is an important resource needed to keep the project progressing on schedule. The dates and amounts of invoices and payments are important. The mathematical difference at any point in time is called the cash flow. It is not uncommon for the cash flow to be negative in that the electrical contractor has spent more money than he/she has been paid. However, the graph in the PDD will clearly show when payments are slow in being made and when only partial payments are being made. Figure 2.6 is a hypothetical example of the graph requested in the PDD. The difference between the expenditures and receipts curve, called the cash flow, shows the amount of money the electrical contractor has yet to recover from the project. Two graphs are shown. When the cash flow is negative by a significant amount, this is a clear red flag. The cash flow can become negative because the owner only makes partial payments, unanticipated difficulties and delays in the work, or late payments by the owner.

Section 4: Inhibitors (Disruptions)

Two situations are monitored that are signs of inhibitors. While there can be others, only changes and out-of-sequence MEP (mechanical, electrical, and plumbing) work are monitored in the PDD. Both parameters are likely to surface later in the project rather than in the civil phase.

Changes disrupt the flow of the work and often lead to delays in completion. The consequence of numerous changes is sometimes schedule acceleration that leads to out-of-sequence work, scheduled overtime, and overmanning. Research has shown that when the percentage of the work hours spent on change order work compared to the total work hours spent to date exceeds 10 percent, impact damages are likely to occur (Leonard 1987). The PDD asks the user to estimate the electrical work hours spent on changes and rework and compare this number to the total hours charged. A graph similar to the hypothetical one in Figure 2.7 should give an indication when rework and changes are problematic. The hours

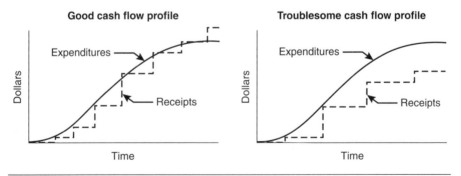

Figure 2.6 Hypothetical cash flow curves

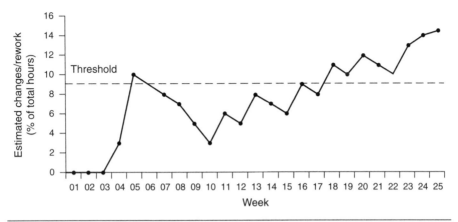

Figure 2.7 Hypothetical change order work hour curve

spent on changes and rework need not be the result of an owner-directed change as the owner may refuse a change directive or be slow in processing the change. The figure shows that beginning with week 11 there is a steady increase in the changes of work hours. As of week 18, more than 10 percent of the total work hours is being spent on changes and rework. This trend is a cause for alarm for the electrical contractor. Only the changes/rework hours of the electrical contractor are tracked. Significant changes performed by others should be noted in Section 1 (Key Events) of the PDD.

Summary of Early Warning Signs

From the information contained in the PDD, a number of early warning signs are evident. The major ones are summarized:

- The likelihood of a time extension was remote.
- The storm drainage was subjected to numerous design changes and was 12 weeks late in being completed.
- Many unforeseen underground utilities were encountered.
- As of week 25, the major activities were at least four weeks behind schedule.
- The design of the foundation was complex and necessitated that working drawings be prepared. The work took nine weeks to complete instead of the four weeks planned. Changes in the structural steel submittals were required.
- The electrical contractor's labor utilization has deviated from the plan.
- There were no required dates for submittals to be furnished.
- The submittal review process is slow.

Overall, many of the problems on Case Study 1 were caused by design errors. Additionally, unforeseen utility locations slowed the storm drainage construction and footing excavation. While beyond the control of the general contractor and subcontractors, there are many ways to respond, and the response of the general contractor is examined in the appraisal.

Appraisal

An assessment of the project was made at the end of week 32. The purpose of this assessment is to determine if the early warning signs disappeared or perpetuated themselves in other ways. Generally, effective management can overcome specific events such as the storm sewer delay and the complex foundation design. However, where contractor management is weak or ineffective, the problems encountered will perpetuate themselves in other ways. Therefore, it is important to be sensitive to management deficiencies as opposed to unforeseen events.

A review of the milestone schedule in Figure 2.8 shows that the HVAC/plumbing rough-in started more than eight weeks late and the electrical rough-in has yet to start. It is more than nine weeks late. Clearly, the problems with schedule delays worsened after week 25. This is illustrated in Figure 2.9. One of the possible causes of the additional schedule slippage is that there were delays in getting the coordination drawings approved by the owner. The reason was probably that the coordination drawings were submitted to the owner late. What appears to have happened is that the general contractor was slow initiating work on the drawings as is evident by the 4½-week delay between the completion of the structural steel and the start of the plumbing rough-in. It is known that the subcontractors were pleading for these drawings well in advance of their approval by the owner.

There are indications that the general contractor did not respond well to the delays in the schedule. There were at least three opportunities to shorten the schedule. They are as follows:

1. The general contractor was slow to recognize the need for working drawings of the foundation. If these had been prepared prior to the start of the foundation, a potential schedule reduction of four weeks may have been realized.
2. The structural steel erection was planned for six weeks. It was done in two phases with one truck-mounted crane. If a second crane and crew was used, the erection time could have been reduced by up to three weeks.
3. The general contractor did not prepare coordination drawings in sufficient time to avoid delays to the subcontractors. This resulted in a delay of up to 4½ weeks.

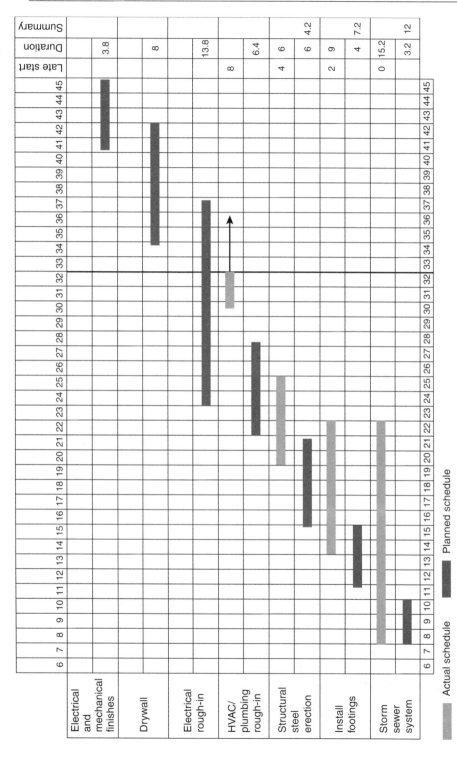

Figure 2.8 Milestone schedule for Case Study 1

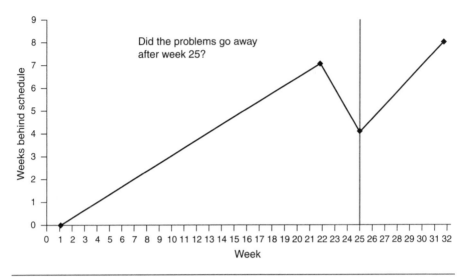

Figure 2.9 Summary of schedule delays for Case Study 1

Overall, there was a potential to reduce the schedule by up to 11½ weeks. The principal deficiency in the general contractor management was that the planning horizon was not large enough to allow for advance planning.

The warning signs could be divided into two types: (1) management issues and (2) unforeseen (or one-time) events. Generally, unforeseen events can be overcome, but management deficiencies will manifest themselves in other ways as the project evolves. In Case Study 1, both types of warning signs were observed as follows:

- Management issues
  - Less than adequate information management
  - Marginal housekeeping practices
  - Short planning horizon
- Unforeseen or one-time events
  - Storm drain design changes
  - Unforeseen underground utilities
  - Complexity of the foundation work

The general contractor did not respond positively to the one-time events or the difficulties in managing the information resource. Opportunities to shorten (or recover) the project schedule were missed. As a result, schedule acceleration was needed to complete the project on time.

CASE STUDY 2

Case Study 2 was the construction of a three-story, three-building dormitory complex on a small private university campus in central Pennsylvania. Units A and C are two-story, townhouse-type structures. Unit B, the largest of the three, is a two-story (plus a basement), structural steel building. Building B is the primary focus of the PDD case study. The total cost of the project was $2,979,000 and the scheduled duration was for 23 weeks. A time extension was unlikely, as the buildings were needed for the upcoming academic year.

Two evaluations are reported herein. The first evaluation is made at the end of week 12, when the project was in the midst of the civil phase. The first 12 weeks are referred to as the documentation phase. The potential problems noted at that time are summarized. Thereafter, at the end of week 20, an appraisal is made regarding whether the conditions noted at the end of week 12 disappeared or worsened. The two assessments simulate possible visits by home office managers and executives.

Documentation Phase

The PDD is generally organized according to Figure 2.1. Other relevant information is also included. The PDD contains four sections as described in the following paragraphs.

Section 1: Project Participants and Key Events

On Case Study 2, the design-bid-build delivery system was used. The general contractor negotiated a cost reimbursable guaranteed maximum price (GMP) contract. The subcontractors had lump sum contracts with the general contractor. The principal participants and contract amounts were recorded and raised no particular red flags. The key events that were recorded included the following:

- At the direction of the owner, the excavation was started in the parking lot instead of starting with the building footprint
- There were labor inefficiencies in setting anchor bolts and bearing plates; the cause was not readily apparent
- Roofing subcontractor was terminated after having done the Unit A roof because the workmanship was unacceptable
- The general contractor did not specify required submission dates for submittals provided by the subcontractors

The last two entries are particularly troublesome because they suggest general contractor management issues that may not be easily overcome. One-time events like the first entry are less bothersome. Thus, early in the project, there are signs that warrant close scrutiny on a weekly basis by the electrical contractor.

Site photographs showed poor housekeeping, which is an indicator of a poorly managed site. It is unlikely that one would observe poor housekeeping on a well-managed site. Thus, poor housekeeping is an important red flag.

Section 2: Progress

Figure 2.10 shows a milestone schedule as of week 12. As can be seen, the footings and piers took five weeks instead of the two weeks planned. Undoubtedly, the inefficiencies in installing the anchor bolts and bearing plates contributed to this delay. The structural steel is nearing completion and is about a week late. The roof

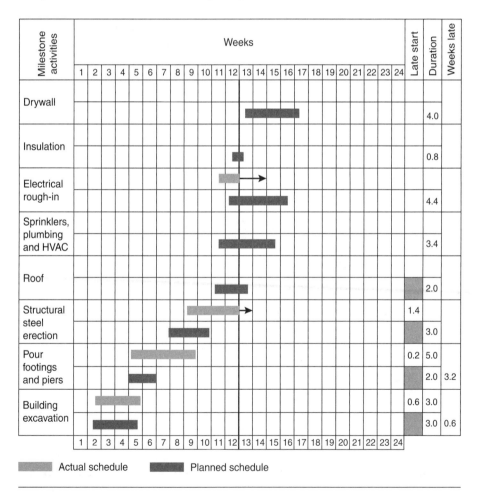

Figure 2.10 Milestone schedule of civil work for Case Study 2

has yet to start (the original roofing subcontractor was terminated) and is nearly two weeks late. The electrical rough-in has begun, but the sprinkler, plumbing, and HVAC rough-in has not.

Overall, the schedule shows signs of being delayed. The early delays to the footings and structural steel may be recovered, but it will be difficult given such a short overall schedule. There are signs that the interior work may have been started out of sequence.

Section 3: Resources

The major resource providing an early warning sign on the case study project is information. Figure 2.11 shows the key submittals that were tracked. The most obvious indicator of distress is that there are no required submission dates for submittals. This is a significant deficiency in the management of the project. The second line in the schedule shows signs of distress in the anchor bolt submittal. The submittal was approved on April 6 and the work began on May 14, some 24 days later than planned. Late submittals might have been the cause of the delays with the piers and footings (see Figure 2.10).

The structural steel submittals appear to have been timely, so any delays to the steel erection would likely have been for other reasons. The elevator submittal was not approved until three days after the work was supposed to have started. The actual installation was delayed by 71 days.

Section 4: Inhibitors (Disruptions)

There were no changes of any significance on the project. Also, the interior work had not begun as of week 12, so it was not possible to track the sequence of the work.

Summary of Early Warning Signs

There are a number of early warning signs on this case study project. It should be quite clear that this project was affected differently than Case Study 1. The early warning signs are:

- The likelihood of a time extension was remote.
- Overall, the project is three to five weeks behind schedule.
- The submittal schedule shows signs that some submittals may have been late and may have delayed the footing and pier activity. The submittal process needs more management attention to avoid further delays.
- There are early indications that the finishing and service trades may have initiated their work out of sequence.

Submittals ID	Original		Re-submission (3)	Final approval		Activity start date		Submittal lateness (5–4)	Activity lateness (6–7) days
	Scheduled (1)	Actual (2)		Scheduled (4)	Actual (5)	Planned (6)	Actual (7)		
Rebar shop drawings (unit B)	NS	3/22/XX		NS	3/30/XX	4/23/XX	4/03/XX	NS	20
Anchor bolts	NS	3/29/XX		NS	4/06/XX	4/20/XX	5/14/XX	NS	(–)24
Structural steel shop drawings JS1, E1, 1 thru 17	NS	3/29/XX		NS	4/06/XX	5/01/XX	5/01/XX	NS	0
Column and beam details E1, E2, 18–39 and 41–48	NS	4/05/XX		NS	4/11/XX	5/01/XX	5/02/XX	NS	(–)1
Structural beam detail E2 and 49 thru 51	NS	4/09/XX		NS	4/17/XX	5/01/XX	5/03/XX	NS	(–)2
Steel joists shop drawings J1 and J2 (sealed)	NS	3/29/XX		NS	4/06/XX	5/01/XX	5/04/XX	NS	(–)3
Steel deck shop drawings D1 thru D3	NS	4/02/XX		NS	4/06/XX	5/01/XX	5/04/XX	NS	(–)3
Prefabricated wood trusses shop drawings (unit B)	NS	5/07/XX		NS	5/23/XX	5/16/XX	6/05/XX	NS	(–)11
Final elevator shop draw/calculations	NS	5/07/XX		NS	5/18/XX	5/15/XX	7/26/XX	NS	(–)71
Sprinkler piping layout dwgs FP1 of 5 thru FP5 of 5	NS			NS	5/29/XX	5/07/XX	5/01/XX	NS	(–)6
Plumbing shop drawings	NS			NS		5/07/XX	5/01/XX	NS	(–)6
Chiller, fan coil units, blower coil AHU data	NS			NS		5/10/XX	5/10/XX	NS	0
Electrical drawings ES-1, E-1 through E-13	NS			NS		4/30/XX	4/30/XX	NS	0

NS—no scheduled date

Footings, roof, and elevator may have been affected by late submittals

Figure 2.11 Submittal schedule for Case Study 2

- The management of subcontractors, as evident by the poor quality work by the roofing subcontractor (and subsequent termination), possible out-of-sequence work, and the deficient housekeeping practices, is an aspect of the project that deserves careful monitoring.

Overall, the early warning signs summarized here suggest that the general contractor management on the project is less than desirable. Information management and subcontractor management are two areas of concern.

Appraisal

An appraisal was performed on Case Study 2 at the end of week 20. The purpose of this appraisal is to determine if the warning signs noted were an aberration of if they were truly warning signs of impending gloom.

In Case Study 2, most of the warning signs are management issues:

- Management issues
  - Less than adequate information management
  - Passive subcontractor management
- Unforeseen or one-time events
  - At the direction of the owner, excavation began in the parking lot instead of on the building footprint

The management issues will be difficult to overcome in Case Study 2. The appraisal explores this situation. Several additional events occurred after week 12 that are worth mentioning. These are:

- The roof was late being constructed. The finish and services trades began their work before the roof was finished. Rain caused damage to the insulation on the second floor.
- Much of the finish work shifted from the second to the first floor, leaving incomplete work.
- The insulation subcontractor was extremely late starting the work.

The marginal housekeeping practices by the subcontractors continued after week 12. Several photographs showed other difficulties in the work.

Figure 2.12 shows the milestone schedule as of week 20. The insulation and drywall work is late; the work is as much as 7½ weeks behind schedule. The delays caused by the footings and piers and the roof have been followed by delays to the finish trades. As shown in Figure 2.13, the schedule delays did not go away, but instead worsened.

The problems caused by the incomplete roof (and subsequent rain) had significant consequences on the project. The schedule delays exacerbated the problems. The general contractor responded by directing work to be done out of

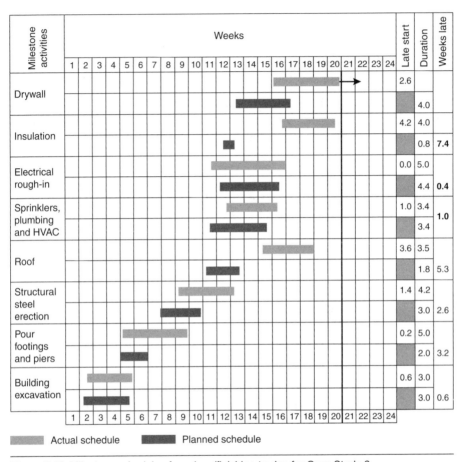

Milestone activities	Weeks																								Late start	Duration	Weeks late
	1	2	3	4	5	6	7	8	9	10	11	12	13	14	15	16	17	18	19	20	21	22	23	24			
Drywall																									2.6		
																										4.0	
Insulation																									4.2	4.0	
																										0.8	7.4
Electrical rough-in																									0.0	5.0	
																										4.4	0.4
Sprinklers, plumbing and HVAC																									1.0	3.4	1.0
																										3.4	
Roof																									3.6	3.5	
																										1.8	5.3
Structural steel erection																									1.4	4.2	
																										3.0	2.6
Pour footings and piers																									0.2	5.0	
																										2.0	3.2
Building excavation																									0.6	3.0	
																										3.0	0.6
	1	2	3	4	5	6	7	8	9	10	11	12	13	14	15	16	17	18	19	20	21	22	23	24			

▨ Actual schedule   ▮ Planned schedule

Figure 2.12 Milestone schedule of services/finishing trades for Case Study 2

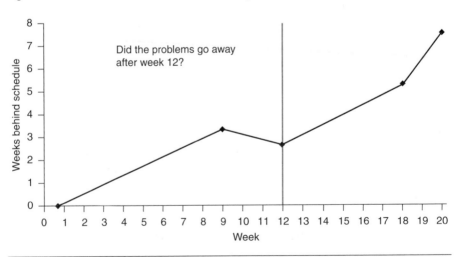

Did the problems go away after week 12?

Figure 2.13 Summary of schedule delays for Case Study 2

sequence. This is illustrated in Figure 2.14. The major services and finish work is shown along with the planned and actual work location. The solid line shows when each trade was scheduled to begin irrespective of the floor. The figure shows an orderly plan with one trade following another. The dashed line shows how the work actually began. Notice that the orderly trade progression was not followed. Also, several trades started on the second floor as planned, but other trades started on the first floor.

In conclusion, the project suffered from weaknesses in project management so that the warning signs observed early in the project did not go away. Overtime and overmanning by the subcontractors was required to complete the project on time. There were two major management deficiencies noted:

1. Information management
   - No required submittal submission dates
   - Late submittals
2. Subcontractor management
   - Roofing subcontractor had to be terminated, which delayed the roof on Building B
   - Steel erection subcontractor took longer than planned
   - Coordination and sequencing of trades was inadequate
   - Unsatisfactory housekeeping by subcontractors
   - Insulation subcontractor was slow responding because he/she was working on another project

CONCLUSIONS

Two case studies were used to illustrate the value of using the PDD. Two evaluations were made on each project. One occurred at the end of a documentation phase, which covered most of the civil construction phase. This evaluation involved the documentation of early warning signs. The second evaluation occurred after the service and finish trades had started work. This appraisal was intended to assess if the early warning signs had disappeared or had evolved into more serious forms of distress. The two evaluations are representative of site visits by home office managers and executives.

The early warning signs at the first evaluation were arranged into two categories: (1) management issues and (2) one-time events. The case studies provided examples of both. It was hypothesized that in the absence of management deficiencies, the one-time events that impacted the project could be overcome with effective general contractor management and that the schedule delays could be recovered. The case studies did not allow the testing of this hypothesis.

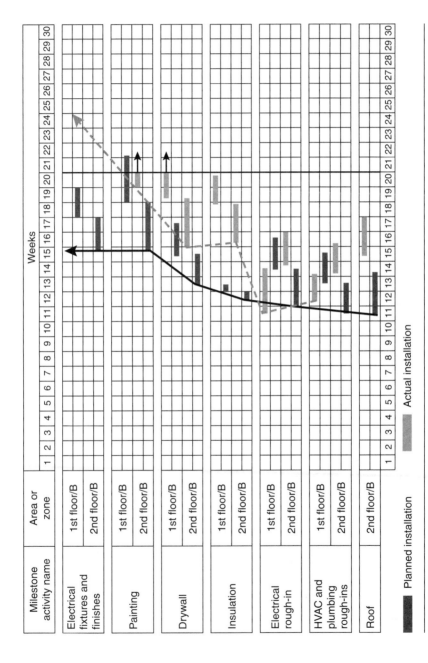

Figure 2.14 Work sequence schedule of MEP trades for Case Study 2

Conversely, management deficiencies do not support easy schedule recovery. Instead, management deficiencies during the documentation phase will likely manifest themselves in other ways and were still evident during the final evaluation on both case studies. These schedule impacts were ultimately overcome by using scheduled overtime and overmanning. This is the classic schedule acceleration scenario, although not every project that experiences schedule acceleration has management deficiencies as the root cause.

The PDD showed that it is possible to track limited, but targeted information and expose red flags or early warning signs of project distress. The information is easy to gather and decipher. Most information is obtained from the project schedule and weekly or biweekly progress meetings. Most information can be readily obtained and updated by a clerk on the site. Weekly updates will highlight areas of concern in sufficient time for corrective action to be initiated.

WEB ADDED VALUE
PDD.pdf

This file provides a blank copy of the PDD used in the sample case studies. Even though this PDD sample has been created for electrical contractors, it is easily customizable for use by other specialty contractors as guidance for creating their own forms and tools to implement the PDD in their projects.

Web
Added
Value™

This book has free material available for download from the
Web Added Value™ resource center at *www.jrosspub.com*

CUMULATIVE IMPACT OF CHANGE ORDERS

Dr. Awad Hanna, *University of Wisconsin–Madison*
Dr. Jeffrey Russell, *University of Wisconsin–Madison*

INTRODUCTION

Change is inevitable on construction projects, primarily because of the uniqueness of each project and the limited resources of time and money that can be spent on planning, executing, and delivering the project. Change clauses, which authorize the owner to alter work performed by the contractor, are included in most construction contracts and provide a mechanism for equitable adjustment to the contract price and duration. Even so, owners and contractors do not always agree on the adjusted contract price or the time it will take to incorporate the change. What is needed is a method to quantify the impact that the adjustments required by the change will have on the changed and unchanged work.

Owners and our legal system recognize that contractors have a right to an adjustment in contract price for owner changes, including the cost associated for materials, labor, lost profit, and increased overhead due to changes. A more complex issue is that of determining the cumulative impact that single or multiple change orders may have over the life of a project.

This chapter introduces two models developed through a research project conducted under the sponsorship of the Construction Industry Institute (CII), ELECTRI International: The Foundation for Electrical Construction Inc. (EI), and the Mechanical Contracting Foundation (MCF). The first model identifies if a construction project has been impacted as a result of cumulative change, while the second predicts the probable magnitude of the cumulative impact due to that change.

FUNDAMENTAL CONCEPTS

Many courts and administrative boards recognize that there is cumulative impact above and beyond the change itself. However, current construction contracts do not typically include adequate language to enable fair and equitable compensation for the unforeseen impact of cumulative change. Often, the contractor fails to foresee, and the owner fails to acknowledge, the "synergistic effect" of the changes on the work as a whole when pricing individual changes. That is, a change order for one area of the project may likely affect other areas of the project as well as the specific area where the change will actually occur. Consequently, projects that exceed cost or schedule targets are likely to lead to claims. Determining the impact that changes can have on contract price and time can be arduous due to the interconnected nature of the construction work and the difficulty in isolating factors to quantify them. As a result, it is difficult for owners and contractors to agree on equitable adjustments, especially for cumulative impact.

The Delta Approach

To study the productivity loss associated with change orders, a standard method to calculate the productivity for a project is needed. The method needs to represent the cumulative impact of changes including any ripple effects. To do this we have defined the term *Delta*, which is shown graphically in Figure 3.1 as the difference between total actual work hours expended to complete the project and the estimated work hours (including the approved change order hours).

Positive values of Delta indicate that the actual productivity is less than the planned or estimated productivity. Negative values of Delta are an indication of higher productivity than originally planned or estimated. Delta can be attributed to a contractor's incorrect estimate, a contractor's inefficiency, or the impact of productivity-related factors such as change orders, weather conditions, work interruptions, and rework among other factors. To study the impact of change, we solicited projects where change orders were thought to be the main reason for loss of efficiency and not other factors such as inaccurate estimates or weather, the Delta then represents the effect of the change orders on the project. We reviewed the data submitted and determined this to be the case.

To compare projects of varying size, it is necessary to scale Delta by the project size, which yields the term *Percent Delta* (%Delta). %Delta is defined in

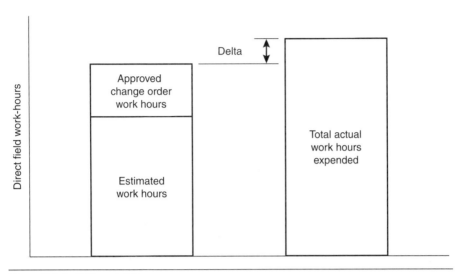

Figure 3.1 Definition of delta

Equation 3.1 as Delta divided by the actual work hours utilized to complete the project. The formal definition of %Delta is as follows:

$$\%\text{Delta} = \tag{3.1}$$
$$\left(\frac{(\text{Total actual direct labor hours}) - (\text{Estimated hours} + \text{approved change order hours})}{\text{Total actual direct labor hours}}\right)(100)$$

The Delta approach allows for a macro analysis that accounts for the full ripple effect that change orders have on the entire project.

Methodology

Our research team consisted of six representatives from CII and three representatives each from EI and MCF. Team members developed a draft questionnaire to provide data on factors that the team felt could, based on experience, influence change orders during execution of a construction project. A pilot study was then conducted to determine how easily the draft questionnaire could be answered and if the data collected would be useful in achieving the research objectives. After reviewing the data and talking to the participating contractors, the research team developed the final questionnaire.

The first section of the questionnaire contained general background questions about the contractors. The other two sections considered an "on budget" project and an "over budget" project. Contractors were requested to submit projects that were perceived to be over budget as a result of change orders, not factors such as low estimates, unforeseen weather conditions, or poor field planning. The projects were investigated based purely on work hours and not by cost (dollars) because work hours are directly comparable. Dollars add complexity for a number of reasons, among them pay scale, premium time differentials, and material costs. In addition to the questionnaire, actual and estimated manpower-loading curves or weekly labor hours were requested for each project along with the change order log, if available.

IS CUMULATIVE IMPACT REAL?

Owners and contractors seldom agree on the existence of the cumulative effect of change orders, let alone to what degree they exert a loss of productivity on the so-called unaffected scope of work. Owners typically require contractors to provide hard evidence that they were affected in areas other than the scope dealing with the change order. Contractors have found that damages resulting from cumulative impact are difficult to calculate and even more difficult to recover. Many attempts to prove that this phenomenon exists have resulted in litigation. Courts and administrative boards recognize cumulative impact, and a number of court decisions have set the standard in terms of what contractors need to provide in the way of convincing evidence. Sheer magnitude of change does not guarantee that the courts will award damages for impact. The contractor seeking to recover for cumulative impact has the burden of providing the proof of cause and damage. The following examples demonstrate this principle.

Case Study Examples

In the first example, the Board of Contract Appeals denied the contractor's claim for cumulative impact based on 202 change items because the contractor was unable to show that the change orders caused any disturbance in the work. According to the Board, the contractor could not meet its burden "by merely offering proof of the issuance of the change requests."

In the second example, a contractor was performing modifications to a launch platform. During the project, the contractor sent the owner a large volume of Requests for Information (RFIs) as a result of various design deficiencies on the project.

The contractor claimed that the number of RFIs was far greater than it could reasonably anticipate, and that the necessity of writing RFIs and waiting for answers reduced its labor productivity. The Board found that the owner was liable for the disruptive effect of the many RFIs, but that the contractor still had the burden of proving its losses. The Board stated that such a burden could be met only through contemporaneous project documentation that recorded and measured the disruptions. As a result, the Board would not allow the contractor to use the total cost approach to its claim. However, the Board did award 10 percent of the amount claimed by the contractor. This leaves us to conclude that if the contractor had provided better-documented evidence of the effect of changes issued by the owner, it could have established entitlement to the full amount of the claim.

In another case, the contractor was able to show through a preponderance of evidence that the cumulative effect of change orders did cause loss of labor productivity. A project owner issued a series of 26 change orders to correct defective drawings and specifications. In addition to direct costs, each change order included 15 percent overhead for indirect costs. The Board ruled that this markup did not adequately compensate the contractor for the cumulative disruption of the performance and administration of the contract. The contractor was awarded additional compensation for cumulative impact costs. This expressly recognized the cumulative impact costs and the contractor was awarded lost labor efficiency.

Mechanical and Electrical Contractor Data

For this research, 33 mechanical contractors responded with data on 65 projects, of which 57 met the criteria of lump sum projects and therefore were entered into our database. Similarly, 35 electrical contractors replied with data on 75 projects, of which 59 were within the limits of our model and therefore were entered into our database. For statistical analysis purposes, projects termed *industrial* include industrial, manufacturing, wastewater treatment, and power plant projects.

Of the 57 mechanical projects, 58 percent were industrial, 33 percent were institutional, 7 percent were commercial, and 2 percent were residential. Of the 59 electrical projects, 50 percent were commercial, 29 percent were industrial, 17 percent were institutional, and 4 percent were residential.

Findings

Characteristics of projects that are impacted by change were identified through hypothesis testing and analysis of variance techniques. Based on the collective experience of our research team, we developed a list of these characteristics or "influencing factors" that we felt led to change order impacts. These factors, shown in Table 3.1, were grouped by degree of impact and whether they were pre-award or post-award considerations.

While administration boards and courts recognize that there is such a phenomenon as a cumulative impact or ripple effect, they require that contractors show a preponderance of evidence to substantiate their claims. Our list of influencing factors was tested against our hypothesis in each of the 116 projects in our database. Based on the research data, we concluded that percent change is a major contributor to producing an impacted project.

Table 3.1 Initial hypotheses of factors that influence change order impact

Pre-Award factors	Post-Award factors
High impact	
Project size	Percent of change
Estimated manpower loading	Timing of change
Quality of estimate	Quality of preplanning
Bid document rating	Materials management
Schedule-driven project	Schedule compression
Renovation work	Unknown conditions
Percent design complete	Lead time
Operating unit	Allowance for extension
	Stacking of trades
	Effectiveness of team
Medium impact	
Original duration	Tools and equipment
Type of project	Availability of manpower
On-site project management	Weather
Cost-driven project	Project control management
Public or private	Material handling constraint
New or repeat project	Manpower density
Constructability review	Craft turnover
Relationship with owner	Experience with owner
Experience with owner	
Local/Remote project	
Owner-furnished equipment	
Low impact	
Delivery system	Closeout and turnover
Contract type	

DETERMINING IMPACT

Using the data from the questionnaires, we applied logistic regression techniques to test the influencing factors to develop a model that would predict the probability that a project has been impacted by change orders. Before any of the statistical analysis could be completed, a preliminary definition of an impacted project had to be developed. The contractors classified projects as either "on budget" or "over budget" when they submitted the questionnaires. For our analysis, we first considered "over budget" projects as impacted by change orders and "on budget projects" as not impacted. However, a conservative evaluation of a contractor's estimating ability is +/−5 percent of project labor hours, so this was set as the lower limit of %Delta for impacted projects. The initial contractor classifications of projects as impacted and not impacted were then compared to a cutoff line drawn at +/− 5 %Delta. Impacted projects with a %Delta of less than 5 percent were reclassified as not impacted. Only two mechanical and six electrical projects were reclassified based on these criteria.

The influencing factors shown in Table 3.2 along with their related coefficients and p-values are the result of the logistic regression.

Figure 3.2 is a graph that shows the importance of p-values. The p-value is a statistical measure indicating the significance of the influencing factors.

The variables on Table 3.2 make up the model or tool required to determine if a project has been impacted. The actual tool can be simplified for use as shown in Equation 3.2.

$$\text{Probability} = \frac{e^x}{1+e^x} \tag{3.2}$$

Table 3.2 Logistic regression results

Influencing Factor	Coefficient	P-value
Constant	−7.00	0.000
Mechanical or electrical*	−1.09	0.143
Percent change X mechanical or electrical	3.89	0.006
Estimated/Actual peak manpower	−1.04	0.039
Processing time	0.63	0.002
Overmanning	2.64	0.000
Overtime	1.19	0.052
Peak manpower ÷ Average manpower	1.20	0.036
Percent change orders related to design issues	0.02	0.074

*Mechanical or electrical is a binary (indicator) variable that takes a value of "1" if the project is mechanical and "0" if the project is electrical.

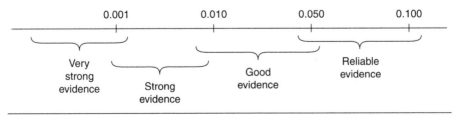

Figure 3.2 Relative strength of p-values

This simplified version will be used in the following example to demonstrate how to determine if a project was impacted. The values to be used for the variables in Table 3.2 come directly from the example project.

Example 3.1: Tom, a project manager for Badger Mechanical, has just finished the job summary paperwork for an office building in Anytown, USA. Tom knew

Table 3.3 Sample calculation of the probability of a project being impacted

Factor (1)	Value (2)	Coefficient (3)	Product (4)
1. Constant	1	−7.00	−7.00
2. Mechanical or electrical 1 if mechanical, 0 if electrical	1	−1.09	−1.09
3. Percent change X mechanical or electrical Percent change as a decimal multiplied by 1 or 0	0.55	3.89	2.14
4. Estimated/Actual peak manpower Estimated peak manpower = 40 Actual peak manpower = 50 Estimated/Actual peak manpower = 0.8	0.8	−1.04	−0.83
5. Processing time Average period of time between initiation or change order and owner's approval 1–7 days = 1　　8–14 days = 2 15–21 days = 3　22–28 days = 4 >28 days = 5	5	0.63	3.15
6. Overmanning (Est. peak ÷ Act. peak manpower) <0.77 1 if overmanning occurred on project (<0.77), otherwise 0	0	2.64	0.00
7. Overtime 1 if overtime used to complete change order work on project, otherwise 0	1	1.19	1.19
8. Actual peak manpower/Actual average manpower Actual peak manpower = 50 Actual average manpower = 22 Therefore actual peak ÷ Actual average Manpower = 2.3	2.3	1.20	2.76
9. Percent Change order hours related to design changes and errors Percent design related change order hours as a percent	30	0.02	0.60
			X = 0.92

the job exceeded the original labor hours planned, but did not realize it was quite this much. He asked to be reimbursed for the 3735 hours that were lost due to labor inefficiencies created by the numerous changes. The owner stated that he believed the contractor mismanaged the job and that was the reason for lost labor hours. Tom explains that change orders accounted for 55 percent of the total labor hours on the project and that it took more than 28 days on average to process these change orders. Around 30 percent of these changes were related to design changes and errors. The actual peak manpower was 50, the estimated peak manpower was 40, and the actual average manpower was 22. Tom added that while he did not have to overman his crews, he did have to work them overtime to finish the project on schedule. The owner agreed that changes were a hassle, but did not think that change orders were the main reason for the impact. After a short discussion, the two agreed to use the logistic regression equation introduced in this chapter.

To determine if the project was impacted, Tom must insert the "values" from his project in column 2 of Table 3.3, and then multiply those values by the coefficients shown in column 3 (these coefficients come from Table 3.2), and include the results in column 4. Then, he must add the resulting products to determine the value of x.

Table 3.3 shows that $x = 0.92$. When this value is inserted in Equation 3.2, we obtain a probability of 0.72. Figure 3.3 graphically depicts how to interpret this value. As illustrated, there is ample evidence of impact.

Findings

Based on the data, we have concluded that our model will predict, with a confidence greater than 80 percent, whether a project has been impacted. We are confident that the data used do not contain unreasonable bids or other contractor failings that would have skewed the results. The model utilizes variables that we felt were more intuitive and that tend to be objective in nature. However, while this tool is available and has a high degree of accuracy, there is no substitute for detailed work-hour data (e.g., cause and effect analysis) gathered from the specific project to demonstrate whether a project has been impacted.

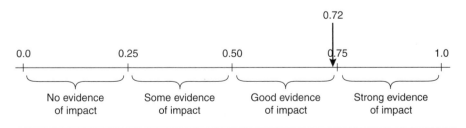

Figure 3.3 Confidence of impacted project

QUANTIFYING IMPACT

Quantifying the impact of change orders involved a two-phase approach. First, hypothesis testing was performed on the influencing variables in the data set containing impacted projects to identify factors that significantly affect productivity loss (%Delta). As a result of the significance testing, six factors integral to the construction industry were identified that directly correlate to %Delta.

The second phase of the analysis required the use of linear regression. The multiple linear regression process involved the development of a model from the independent variables in the database that would predict the value of the dependent variable, %Delta.

Model and Validation

For our final model (Equation 3.3), a total of six variables were identified. The linear regression equation to predict the magnitude of impact of change orders on labor productivity is:

$$\%Delta = 0.37 + 0.12A - 0.08B - 0.17C - 0.09D - 0.05E + 0.02F \quad (3.3)$$

where

  A: Percent change
  B: Percentage time of project manager on project
  C: Percentage of owner-initiated change orders
  D: Productivity variable
  E: Overmanning variable
  F: Processing time variable

Table 3.4 provides the definition for each of the independent factors listed in Equation 3.3. The number in column 3 of Table 3.4 should be considered the limits for the variables in the model. Projects with variables that fall outside these limits lessen the accuracy of the %Delta calculation.

This model was validated through cross-validation and with new data. For cross-validation, the data were divided randomly into five subsets. The model was refitted using approximately 80 percent of the data.

It is important to look at how well the full model would predict %Delta for new data. Seven new projects were used for the validation of the linear regression model. These seven projects were solicited after the close of the data collection and were not considered when the linear regression model was developed. The final results of the validation are shown in Table 3.5.

All seven projects are predicted within +/− 15 percent of the actual %Delta, five projects are within +/− 10 of the actual %Delta, and four projects are within

Table 3.4 Definitions of factors included in Equation 3.3

Factor (1)	Definition (2)	Limits (3)
Percent change (Variable A)	Percent of change on project in terms of original budgeted work hours	2.5% to 90%
Percentage time of project Manager on project (Variable B)	Percent of time the project manager spends on the project	0% to 100%
Percentage of owner-initiated Change orders (Variable C)	Percent of change orders initiated by the owner	0% to 100%
Productivity (Variable D)	Did you track productivity for the project? (input [work hours] ÷ output [units installed])	
	The contractor could use one of the following:	0= No 1 = Yes
	−Track % complete by earned value	
	−Track % complete by actual earned work hours	
	−Track % complete by actual installed quantities	
Overmanning (Variable E)	Did overmanning occur on the project? (Estimated peak manpower ÷ Actual peak manpower) <0.77	0= No 1 = Yes
Processing time (Variable F)	The period of time between initiation of the change order and the owner's approval of the change order:	1 to 5
	1–7 days = 1 8–14 days = 2	
	15–21 days = 3 22–28 days = 4	
	>28 days = 5	

Table 3.5 Predicted %Delta calculations for new projects

Project number	Predicted Delta	Actual Delta	Delta deviation
1	0.1261	0.1575	0.0315
2	0.1410	0.1797	0.0387
3	0.2609	0.1085	−0.1525
4	0.3601	0.2482	−0.1118
5	0.3115	0.3577	0.0462
6	0.4776	0.4395	−0.0381
7	0.1531	0.2533	0.1001

$+/-$ 5 percent of the actual %Delta. To demonstrate how Equation 3.3 should be utilized, the following scenario is presented.

Example 3.2: Mad Dog Electrical entered into a contract to install the electrical requirements for an industrial building for Wando Ventures. The contractor estimated it would take 7500 work hours to complete the project. Throughout the project, Wando Ventures approved various change orders for 515 work hours (44 percent of all change orders were initiated by the owner). The contractor found it necessary to overman the project to stay on schedule. During the construction phase, the contractor did track productivity via actual/estimated manpower curves. The project manager spent 15 percent of his time working on this project. Average processing time for change orders was 32 days. At the end of the project, Mad Dog Electrical adds up its total labor hours and arrives at 13,000 work hours, a Delta of 4985 work hours or 38.35 percent. The contractor consequently reserved his right to make a claim for the impact that the changes may have had on their labor efficiency. Mad Dog Electrical submitted this amount (13,000 work hours) to Wando Ventures to be reimbursed for the cumulative impact of the changed work. Wando Ventures agreed that the changes may have impacted the project and lowered the labor productivity of the contractor, but not to the extent that Mad Dog Electrical claimed. Because both parties agree that the project was probably impacted, they decided to use the model introduced in this chapter to predict the value of %Delta. Therefore, they insert the project data into Table 3.6 to determine the probable magnitude of impact.

Entering the project information into Equation 3.3 yields:

$$\%Delta = 0.37 + 0.12(0.07) - 0.08(0.15) - 0.17(0.44) - 0.09(1) - 0.05(1) + 0.02(5) \quad (3.4)$$
$$\%Delta = 0.26$$

In other words, the parties find that the %Delta for the project was predicted to be 26 percent of the total actual work hours. This corresponds to 3380 extra work hours that could be attributed to changes, in addition to the 8015 work hours (estimate + approved change orders). While 11,395 work hours is lower than the submitted 13,000 work hours, Wando Ventures is not totally convinced that the number is exactly correct. However, the two parties are able to use the predicted value as a benchmark for negotiations.

Findings

It is important to remember when applying any equation, that to acquire accurate results, you must operate within the limits of the parameters used when developing the equation. The limits of our model (Equation 3.3) are defined in column

Table 3.6 Sample calculation of %Delta

Variables (1)	Coefficient (2)	Value (3)	Product (4)
Constant	0.37	1	0.37
Percent change (Variable A)	0.12	0.07	0.01
Percentage time of project manager on project (Variable B)	−0.08	0.15	−0.01
Percentage of owner-initiated change orders (Variable C)	−0.17	0.44	−0.07
Productivity (Variable D)	−0.09	1	−0.09
Overmanning (Variable E)	−0.05	1	−0.05
Processing time (Variable F)	0.02	5	0.10
		%Delta	0.26

3 of Table 3.4. Utilizing this model, falling outside of these limits may produce inaccurate results.

The results of the significance testing identified some variables that can be monitored from the beginning of the project by both owners and contractors to determine whether they are trending in a direction that could adversely affect productivity. Tracking these factors increases the possibility of reacting to negative trends before they result in unrecoverable losses in productivity. Therefore, these factors can be used in a proactive manner by the project stakeholders as an assessment tool before the conclusion of a project.

RECOMMENDATIONS

Because change orders occur on virtually every construction project, managing them can make or break a project, especially in labor-intensive fields, including mechanical and electrical construction.

As our models were being developed, it became apparent that certain practices, if followed, were most likely to keep a project from being impacted, or would at least indicate that a negative trend was starting. A few of these best practices are listed for both owners and contractors.

Recommendations to Owners

Owners have a significant amount of control over the outcome of a project. Many of the decisions they make early in the project will have impacts on the rest of the project. The following recommendations are provided for owners.

Pre-Project Planning

The most common reasons for change orders are additions, design changes, and design errors. By spending more money on the design up front, the owner could decrease the number of change orders associated with design problems, increase the percent of the design complete prior to the award of the contract, and increase the degree of coordination in the designs between the different trades. These are all factors that were found to be significant in this study.

Change Order Processing Time

The time between the initiation of the change order and the owner's approval should be kept as short as possible. By decreasing the processing time of change orders, the project is less likely to be impacted and to see a smaller productivity loss associated with the change orders. There is a strong correlation between the processing time and the percent of the change order hours that are approved by the owner.

Manpower Data

Owners should require contractors to present a manpower loading curve as a part of their proposal. When the actual work hours are updated on a regular basis and compared to the planned manpower curve as the project progresses, it provides an early warning signal that labor productivity has changed. It could be an indication that labor efficiency has decreased (more actual work hours used than planned), or it may indicate that the contractor is performing more efficiently than planned (fewer actual work hours used than planned).

Recommendations to Contractors

Actions of a contractor can impact a project just as easily as those of an owner. It should come as no surprise that some of the recommendations are uniform for owners and contractors alike. This should be the case considering that everyone on a construction project should be working toward the same goal.

Two of the most important tools of a contractor are (1) the schedule and (2) the manpower loading curve. As with the owner, these tools allow the contractor to track the progress of the project as well as many other factors such as the ratios of Actual Peak over Actual Average Manpower and Estimated Peak over Actual Peak Manpower. They also provide a means for defending a claim after the completion of a project should a disagreement arise.

Whether the contractor updated the schedule during construction showed up as a significant factor in many of the preliminary models developed to quantify the productivity loss. Projects where the schedule was updated regularly had lower productivity losses. Another factor that showed up in the models, as well as in the hypothesis testing, is whether the contractor tracked productivity on the project. Projects in which the contractor tracked productivity had lower productivity losses.

CONCLUSIONS

This chapter illustrates that there is a strong correlation between the amount of change items initiated on a project and some loss of productivity (%Delta). As the number of change items increase, one is more likely to have an impacted project. Two models (equations) were developed. While we have confidence in the developed models, it must be reiterated that it is best to use the results in conjunction with hard data developed from the project such as manpower loading curves and productivity tracking data. It is further suggested that both owners and contractors track estimated manpower loading curves against actual work hours as the project progresses. This will allow proactive steps to be taken to correct negative trends rather than waiting for the end of the project to determine to what extent they may have affected the project. We also recommend that the owner and contractor agree to utilize the developed models for change order conflict resolution prior to signing the contract. It would minimally provide a starting point for negotiations.

4

SEQUENCING GUIDELINES

Dr. David Riley, *The Pennsylvania State University*
Dr. Victor Sanvido, *The Pennsylvania State University*
Bevan Mace, *The Pennsylvania State University*

INTRODUCTION

The development of a construction sequence is often considered a project management function or a detailed scheduling exercise. However, sequencing is rarely discussed in the real context of production. Project management and scheduling are often directed toward tracking progress, an important but separate issue from production planning. This chapter focuses on finding ways to create and maintain a productive work environment for specialty contractors.

It has been noticed that (1) the relationships between sequencing and project management in specialty contractors are not well defined or understood, (2) the possibility of gaining competitive advantage and savings by improving sequencing practices is a key method to improving profitability, (3) there is a widely perceived inability or unwillingness of general contractors and construction managers to develop and implement detailed sequencing plans, (4) the role of sequencing in the estimating process and productivity factors used for project planning is of high interest due to the vague conditions on which these factors are based, and (5) the perpetual acceptance of interference problems and out-of-sequence work as an inevitable feature of building construction projects leads to complacency and isolation during the planning process, when what is needed is proactive thinking.

The case study narratives introduced in this chapter demonstrate highly effective sequencing practices on uniquely planned projects and general sequencing practices on typical building projects. Based on those case studies and interviews, a framework for sequencing rules is developed and a set of recommended practices is assembled. Five key sequencing guidelines are identified for coordination, planning, materials and methods, work assignments, and feedback.

BACKGROUND

Previous research has identified several fundamental principles on which to base the development of construction sequence. Echeverry, Ibbs, and Kim (1990) defined a collection of factors on which activity sequences are based, including physical dependencies, trade interaction, path interference, and code regulations. Construction practitioners use these common dependencies every day to develop critical path method (CPM) schedules. More detailed planning that determines work locations and the flow of trades through building spaces is often needed to direct individual crews. Riley (1994) described the key results of these plans to set building sequences (the order that floors are completed—top down or bottom up) and floor sequences (the order that work is completed in specific areas on each building floor). Riley and Sanvido (1997) identified that projects with sequence problems resulted in more interference for mechanical, electrical, and fire protection trades than all other trades combined. Basic sequencing benchmarks were identified for project sequences based on spatial interaction of structural, enclosure, rough-in, and finish phases.

Riley and Sanvido (1997) organized the observations of successful work practices and space planning techniques into a set of steps to develop an efficient sequence plan for mechanical, electrical, and fire protection trades. These steps are:

- Subdivide floor plan into zones. Suggested zones that may be considered independently are the core, perimeter, and floor areas in between. Large floor areas may be subdivided into halves or quarters.
- Determine the building sequence (order that floors are completed). A top-down sequence is often affective for interior and finish trades on low-rise buildings.
- Determine the floor sequence (order work is completed on each floor). Each zone on a floor may have its own sequence. Specify order in which each zone is completed on each floor.
- Determine work rate for each zone based on the driving activity in that zone. Trades should size crews to meet target work rate.

Additional flexible crews can support frontline crews when needed, and perform noncritical tasks when not needed. Target work rates should be established prior to negotiations with subcontractors.

- Assign material storage areas on floors using reflected ceiling plans as a guide to avoid areas where overhead work is required.

Ballard and Howell (1998) demonstrate the relationship of lean manufacturing principles to construction. The principles of production management that have evolved from this relationship form the basis of specific sequence guidelines for specialty contractors. The fundamental principles of lean construction from which sequence guidelines are formed include:

- Stop the line and halt work to correct systematic problems: perform work only when design is free of defects
- Pulling versus pushing: construction, not design information, should drive resource needs
- One-piece flow: crews should complete as much work as possible on each pass through an area
- Synchronize and align: get crews to work at similar paces
- Transparency: share progress information, feedback, and challenges with everyone

Supply chain management also has a direct impact on project sequencing, as available materials and services must be synchronized for efficient field production. Lean supply recommendations for specialty contractors include:

- Make suppliers and vendors into business partners
- Give specialty contractors design-build responsibility
- Reimburse for cost to get control over production process
- Integrate planning and control systems

Ballard, Howell, and Kartam (1994) discuss the "Last Planner" concept as a mechanism to identify how reliable work assignments can be assed. Howell and Ballard (1995) begin to demonstrate how these concepts can be applied specifically to electrical contractors. Spot checks for planning reliability measurements are used to evaluate case study projects to determine if foremen and project managers of specialty contractors are able to predict the available work for the coming week. The results will be used as indicators of sequence problems.

Lean performance improvement strategies for construction include (Ballard and Howell 1998):

- Minimize the impact of erratic delivery and quality. This can be accomplished by (1) stabilizing work process by maintaining workable backlogs and assigning only from backlogs, (2) eliminating delays for materials, equipment, and crew interference, (3) making the foreman a manager, and (4) getting the facts about delays and rework.

- Reduce cost. This can be achieved by (1) performing constructability reviews, (2) shifting work to shop conditions, (3) structuring on-site work as close as possible to workshop conditions, (4) developing craftsmen with multicraft capabilities, (5) maintaining tools and equipment for zero downtime, (6) improving crew mix on repetitive operations, (7) reducing rework by completing work the first time, identifying and learning from errors, and tracking repetitive errors to their cause, and (8) reducing the number of organizational errors.
- Reduce duration. This can be accomplished by (1) advancing mobilization within the constraints of available backlog and reliable work flow, (2) reducing backlog quantities to reflect more reliable deliveries, and (3) dividing construction sites into subprojects and facilitating functional team approaches.

CASE STUDIES

Several building projects were studied in three geographical regions. The results are presented in cases with unique approaches to planning/sequencing and cases in which the project was "a typical project." On the uniquely planned projects, the techniques that demonstrated effective sequencing methods based on established production management principles were recorded. On the typical planned projects, the planners thought the work was industry standard and did not see the need to plan effectively. In these cases, we documented the typical inefficiencies resulting from informal sequencing practices.

Uniquely Planned Projects and Principles Demonstrated

Medical Research Laboratory in Seattle, Washington

A formal coordination was implemented on the laboratory project. Two-dimensional CAD drawings were generated by each contractor and combined into one file, which was plotted and scrutinized for interference problems. The mechanical contractor also utilized three-dimensional CAD modeling in the mechanical room to coordinate equipment and large pipe spools. A sample weekly schedule for the coordination process is as follows:

- Monday: Each trade submits CAD layers to lead coordinator (general or mechanical contractor).
- Tuesday: Plots of all layers are distributed at weekly job meeting.

- Wednesday: A coordination meeting is held with all contractors and engineers. Items that interfere with each other are marked. Changes are suggested and noted on drawings.
- Thursday, Friday: Drawings are modified and updated based on meeting results. CAD details continue to be generated for remaining spaces.

All trades were strongly encouraged to participate in this process. When an area was coordinated, each trade signed off on the drawing. This demonstrated a commitment to install materials in the correct location, and allowed each trade to take responsibility for the coordinated design.

In the field, the coordination efforts paid off in several visible ways. Since all locations for slab penetrations and hangers were identified, they could be placed as embeds and box-outs on concrete deck forms. This virtually eliminated all core drilling and the installation of hangers was rapid, as the time-consuming layout and installation of hanger inserts did not have to be performed on the ceiling after forms were removed. Another notable feature of the building construction was the breakdown of the slab pours. Instead of placing the floor slab as a series of large pours, daily pours of a manageable size were made on a rigid schedule. Ample time was also allowed for the layout and installation of box-outs and embedded anchors so mechanical and electrical crews did not have to compete with ironworkers for space or work on areas with rebar already installed. The steady pour schedule also equated to a steady stream of work released for crews who were working on top of the deck forms or under the slab as the forms were stripped. This flow rate was used to dictate the pace for all crews. Ceiling hangers were installed almost immediately after form removal, followed by the installation of large ductwork and pipe spools. Although this project included a hanging interstitial space, larger spools of pipe and duct were installed from the concrete floor 18 feet below before the interstitial space was constructed. Once these large items were in place, the hanging steel frame and concrete slab for the interstitial space was built. This allowed the work inside the interstitial space to take place concurrently with the laboratory finishes below.

Finally, a post-occupancy evaluation study of this project revealed an interesting lesson for owners. Although typically viewed as expensive, this interstitial design resulted in several cost savings. Late changes and additions were made with little impact on the project, as space for systems was available. The space provided for maintenance and adjustment of systems also translated into lower overall operation expenses.

Oceanography Laboratory in Seattle, Washington

This project provided an example of how the delays and problems with coordination can be costly to production. On this project, far too little space was left for technical systems, requiring complex, and at times costly, rerouting of systems around congested areas. As a result, certain areas had one system adjacent to the ceiling and in need of sequence priority, while other areas needed different trades to go first. Creating work flow in this situation was nearly impossible. Adding to the pain of the project's superintendent and specialty contractors, the steel frame experienced significant deflection as concrete floors were poured, reducing the height of already crowded ceiling plenums. Items that had been coordinated to fit in these areas now did not have enough space. In some cases, this was not discovered until trades attempted to install work. These problems further delayed the coordination process. When a work area was finally released for construction, all trades tended to move into the area. Crowding was thus significant. It was evident that the architectural and structural designers of this project had a blatant disregard for the space required for mechanical, electrical, and plumbing (MEP) systems, and for the coordination between MEP, architectural, and structural systems.

This project also offered an excellent example of how materials and methods choices could ease work flow in the field. The mechanical contractor proposed a series of alternatives, including factory-assembled batteries of laboratory fittings. The alternative materials were more expensive than individual fittings but required far fewer field connections, thus producing time and overall cost savings. In addition, the ganged fittings were smaller than individual fittings, and thus resulted in space savings in the crowded MEP spaces. Several laboratory pipe systems also required seals made by electrical charge. The system in the specifications required individual seals to be made. The mechanical contractor proposed a similar system that permitted multiple seals to be made simultaneously with parallel connections. This change dramatically reduced the time required to make the seals.

Luxury Apartment in Las Vegas, Nevada

The construction of a five-story, luxury apartment building was evaluated to explore the role of four-dimensional modeling in the detailed sequencing of multiple trades on individual building floors. This project was visited during the structural phase, as multiple trades prepared to follow shoring removal to complete interior rough-in and finishes. The general contractors had prepared a schedule for the work of each trade on a floor at a time, and demanded the completion of one floor every two weeks. No plan for a specific work direction was identified for the floors. The MEP contractors on this project believed they would not be capable of completing the rather large floor plate with the available workforce. In a cooperative meeting between the mechanical, electrical, and drywall contractors (the general contractor elected not to attend this meeting), a more detailed work sequence plan was developed. By carefully reviewing the work on each floor, it was determined that a phased construction of walls would permit open travel lanes for material handling activities necessary after the original scheduled wall construction. Also, by assigning a work direction on each floor, it was possible to manage how crews moved through each floor. This permitted multiple trades to occupy a single floor and remain out of each other's way.

The revised schedule was no longer sequential, allowing the durations of the MEP activities to be extended and more realistic. As the plan was implemented, it was modified slightly to accommodate the shoring removal. However, the trades were able to complete their work with minimal interference problems.

This case provides a classic example of the threshold of current planning and visualization. The contractor was unwilling to schedule more than one trade on any given floor at a time. However, he had not accounted for the fact that several trades did not have the capacity to complete each floor at the stipulated rate. By integrating flow patterns and the phased construction of core and demising walls, it was possible to generate a more feasible production plan. While the duration of the overall floor construction was equal to the original plan, the individual activities were planned at a more realistic pace. It is notable that on the first two floors of this project, the plan was followed and construction was performed on schedule. On subsequent floors, the contractors slipped back into the "free-for-all" approach and ended up having interference problems on floors. The absence of the general contractor from the initial planning, and the resulting inability to enforce the plan, played a significant role in the uncontrolled work environment later in the project.

High-Rise Hotel in Seattle, Washington

On this project, the general contractor allowed the mechanical contractor to take the lead and did not closely manage the coordination process. As a result, there were several instances where other trades felt at the mercy of the mechanical contractor, and left out of the loop. This translated into several delays and interference problems for the electrical contractor. Also, the sprinkler contractor on this project chose not to participate in the coordination process. This was highly evident in the field, as all visible field conflict observed during site visits involved the sprinkler system.

The Sovent® system used on the guestroom floors provided an example of how a more expensive material design actually saved costs by reducing field labor. This system dramatically reduced the required vent piping for the guestroom bathrooms, and permitted large manifolds of the drain waste and vent system to be prefabricated. As wall studs are typically driven by the need to start in-wall plumbing and electrical work, speeding up the in-wall work allowed wall stud construction to be delayed and other work to be performed on the unobstructed floors.

The electrical contractor organized the work on repetitive floors by maintaining established crews to take maximum advantage of the repetitive nature of the work.

Medical Center/Office in Seattle, Washington

This multistory, medical office project included a concrete structure and dense mechanical systems. The coordination on this project was performed before construction. However, problems arose during superstructure construction and the project fell behind schedule. In efforts to catch up, work areas became congested, and out-of-sequence work became a problem.

The coordination effort on this project proved useful at avoiding conflict. However, the process was marred by difficulties with the design and obtaining answers to questions from engineers. The concrete sequence on this project also fell behind schedule for several legitimate reasons. As a result, specialty trades were forced to work on a more condensed schedule, and were found to be crowding each other on floors. MEP contractors stated that the delay of a building's structure, combined with the rigid completion deadlines, was the most common cause of crowding and out-of-sequence work later in the project. An interesting point to note on this project was again the effects of the concrete sequence on the production of ME work. The floor plates on this job were poured in two large sections. As a result, more time was needed between slab pours. This made it difficult to maintain a rhythm as specialty contractors attempted to layout over 700 embedded anchors per floor.

High School in Seattle, Washington

This high school project was studied from the perspective of the electrical contractor, who had developed a unique work sequence plan and foreman information system. An initial work sequence was identified on this project and visible efforts were made by the electrical contractor to stay on the planned sequence, as dedicated crews moved through a series of predefined zones. When it became apparent that the general contractor required help with the electrical work beyond the planned sequence to keep the commissioning process on track, a special crew was formed to respond to this less predictable work. This allowed other crews to progress with fewer interruptions.

The electrical contractor insisted that they be permitted to use large rolling carts for all materials. This requires all work areas to be cleared of debris prior to beginning work in that area. Weekly planning reports also required foremen to report any hindrances to their work, such as incomplete work or trash piles in the way of their carts. These hindrances were provided to the general contractor. This feedback brought the electrical crews' needs to the attention of the general superintendent, who responded by pushing other trades, especially drywall, to keep up with waste removal and other work that was needed before the critical electrical work could proceed.

A unique planning and information tool on this project was developed to summarize design information for foremen. A large notebook was assembled for each work zone, which identified all power system design requirements, in-slab conduit plans, architectural plans, and wall sections. The highly organized format of these notebooks, which cost over $50,000 to assemble, dramatically reduced the time needed for foremen to locate and complete work with their crews. Rework was also reduced, as fewer mistakes were made, and design changes were more readily communicated on pages of the notebook, as opposed to shop drawing addenda. Both the foreman and the project manager of the electrical contractor said that this format reduced the time required to study and clarify design information by at least half, saving over 10 hours per week, per foreman, and easily paying for the investment in the manual.

Rocket Manufacturing Facility in Decatur, Alabama

This project was lead by a design-build joint venture. The main factory originally was designed to be 2.1 million square feet, but following changes by the owner in the rocket manufacturing process, the size was reduced to 1.5 million square feet. When these changes were made, construction was underway and steel had been purchased for the original design. Aside from the impacts of this significant change in size and relocation of equipment, the owner also was to take partial occupancy as the facility was completed. This created additional coordination issues to ensure that the owner was able to use the space already turned over without interruption and that crews could work effectively.

Several unique coordination methods were used on this project. First, a central tunnel 20 feet wide, 20 feet deep, and nearly a mile long was created to house 80 percent of the utilities to create an open manufacturing floor. Space in the tunnel (vertically and horizontally defined) was allocated for each trade, so that the mechanical contractor had specific space to install pipe and sheet metal, and other trades had similar areas. This reduced physical conflicts with the MEP systems. Another unique method was demonstrated by utilizing some of the nearly 33 bridge cranes in the facility. Crews installing materials in the ceiling utilized platforms that were hung from the crane rails. The design-builder scheduled when each trade was to use each platform. A crew would load its materials and tools on the platform in the morning and would be raised to the work area. A typically sized platform was 100 feet by 200 feet and was able to support man lifts. This effectively created new workspace for the crews to utilize and allowed three crews to be working simultaneously in the same footprint area: one in the tunnel, one on the ground floor, and one on the platform.

With the methods just described, timing coordination not physical coordination was identified as the greatest issue on the job. With the degree of design changes, the design team struggled to keep up with the construction efforts, and interferences were left for the crew in the field to identify and resolve. Weekly coordination meetings were led by the design-builder and involved the superintendents from each contractor on the job. MEP coordination and potential interferences were not necessarily discussed, but rather the scheduling of trades into workspaces in the facility became the focus. Individual contractors would have daily meetings between foremen working in the same area to coordinate. Supervisors of some contractors would have daily discussions in the morning on what was planned to be accomplished and then at the end of the day to discuss what had occurred. A reason analysis was completed to explain any variation for future planning.

Typically Planned Projects

Library Expansion in State College, Pennsylvania

This is the main library in a major university and continuous operation was critical to support the university's education and research programs. The expansion included a 100,000 square feet, five-story addition to house new collections as well as administrative office space. The renovation included work in existing collection areas, new circulation paths, and customer support areas, as well as making the library compliant with the Americans with Disabilities Act (ADA).

The project first started by contracting with a national contractor to create a phasing plan for the project. This divided the expansion/renovation into manageable phases and allowed for programming to be done to handle the library's large collection of resources. The contractor who did this phasing plan was not selected as the general contractor for the project.

The general contractor dictated the sequence and provided biweekly schedule updates to the contractors. The general contractor attempted to proactively coordinate, but for the most part ended up reacting to problems in the field and troubleshooting problems. The planned sequence was sheet metal, piping, sprinklers, and then electrical. In the field, the sprinkler contractor moved in after the sheet metal contractor because they were ready and were not affected by some changes that impacted the piping contractor.

The contractors on-site generally felt that the sequence planning on this job met their typical expectations. They expected the general contractor to lead and inform the specialty contractors where to work. They felt that progress on the job was about average. They handled coordination problems by mutual adjustments when interferences arose. Some sequence problems caused by non-MEP trades were the construction of full-height partition walls and placing of the ceiling grid work before the MEP trades were through the space.

Student Union Building Expansion at State College, Pennsylvania

This facility is located in the center of a major university campus and is a critical facility for the university. The expansion included the addition of a new auditorium, art gallery, and cultural center; a three-story atrium with eateries and seating areas; administrative and student offices; and a recreation center. Among the spaces renovated were a ballroom and auditorium, eateries, study areas, offices, and support areas. This project had a high profile and had significant challenges of sequencing and coordination not only from an MEP perspective, but also in terms of taking over spaces and starting up new areas while the building was still occupied and in use.

The general contractor set up and facilitated the biweekly coordination meetings and aided in solving coordination issues in the field. The mechanical contractor started the coordination process, but all contractors in the job were involved the entire time. Each specialty contractor identified the coordination process as invaluable in delivering this difficult project. To ease the process, the project was divided into four quadrants that were subsequently divided into several blocks. This allowed each block to be handled and the coordination approved individually. It was identified that the coordination meeting should take place about two months before work in that area began. The coordination process used required the mechanical contractor to develop CAD drawings and email them to the plumbing contractor. The plumbing contractor overlayed its piping layout and delivered paper drawings to the fire protection contractor, who did not use CAD. The fire protection contractor then marked up these drawings and delivered them to the electrical contractor. After this final step, the trades met to work out any pending issues.

The sequencing of crews generally followed the same flow as the coordination process. The most common method of scheduling and monitoring progress was a two- or three-week look-ahead schedule. The general contractor produced these look-ahead schedules, which were visually verified for progress. Regular meetings and detailed minutes were identified as crucial for effective communication of the schedule and planning purposes.

Football Training Facility in State College, Pennsylvania

This 80,000 square feet, $12 million facility was completed in 11 months. This facility is the new training center for a major college football team. The facility contains strength training rooms and therapeutic facilities, locker rooms, offices for coaches and staff, a 185-seat auditorium and meeting spaces, and support areas. Originally, the schedule was for one year, but due to the addition of a pre-season game in late August, the facility was requested one month earlier.

The mechanical contractor developed its duct drawings and distributed them to the other trades. Each trade would use these drawings as a base plan and mark up where they planned to install their materials. The duct drawings were selected as the basis for coordination since the contractor gave the ductwork precedence due to the size of the material. Afterwards, all the trades would meet and overlay everyone's drawings to look for interferences. The construction manager oversaw this process, but the mechanical contractor was mainly responsible. Weekly coordination meetings were held and conducted by the construction manager.

The work sequence began in the same manner as the coordination process, with the ductwork beginning first and followed by piping and sprinkler lines. The electricians were generally the last trade through the area. An effort was made to use prefabricated materials as much as possible, but problems arose when the prefabricated sections did not fit in the dimensions of the constructed facility. While not terribly complex, the method that this project team used to coordinate and sequence the work ensured that the project was completed on time.

Multisport Facility in State College, Pennsylvania

This 130,000 square feet, $13 million project was completed in 12 months and included the construction of a large clear span, with an arched roof and gable ends. This facility serves as an arena for track and field competitions as well as containing a multisport artificial turf area for basketball and baseball practice. This facility houses the nation's largest hydraulic track; a 200-meter oval track with turns able to be banked up to 47 inches. The facility uses radiant heating, suspended 30 feet above the floor, for the track and multisport area and a variable air volume (VAV) system is used for the office and support area.

Little coordination of the MEP systems was done for this project. This was attributed to the type of facility being constructed. With large amounts of space available, conflicts between trades were not perceived as a problem. Coordination was handled with meetings between the trades as needed. The engineers did not do any preliminary coordination and all sequence problems were left for the field to resolve. The general contractor did manage the coordination and sequencing process. However, some contractors elected not to participate in this effort. Piping was first installed, followed by sheet metal, then electrical work, and finally sprinkler lines. The mechanical contractor used a two-week look-ahead schedule to monitor progress and to keep crews working productively.

While there were minimal coordination issues between the MEP trades, a major problem arose with the electrical trade and coordination with other trades.

The problem arose with conduit attached to the roof deck and structural members in the open area and the method to install it. The electrical contractor proposed an effective solution to solve the problem and hide the conduit, but created problems afterwards since the roofing contractor had to be delayed. The painting contractor had to be called back in a second time after the track had been installed. Special metal bridges had to be constructed to keep the main lifts off the track for the painter to perform work 30 feet above the floor. This resulted in lost profit and delayed completion for the project.

SEQUENCING GUIDELINES

The case studies summarized, as well as numerous interviews, were used to identify what we believe to be the root causes of success or failure for construction project sequencing. The case studies demonstrate that no trade can plan its own work in a vacuum, and that while it is possible to protect one's own interests in some projects, in most projects a productive work sequence requires a team effort to plan and implement. The following areas represent the core sequencing guidelines identified in this study.

Guideline 1: Coordination

Effective coordination techniques are prerequisite to successful sequencing. A timely coordination process is necessary to minimize or eliminate disruptions to fieldwork. Recommendations:

- Perform coordination before the structure is built
- Have the general contractor control this activity
- The best individuals to accomplish this task would have a blend of foremen experience, CAD ability, and be effective communicators

Guideline 2: Planning Sequences

Partnering between trades is necessary to understand each trade's sequencing and materials and methods constraints. Productive sequences are largely unique to each project, and need to be developed as a team. Recommendations:

- Perform sequence planning like partnering
- Develop detailed work breakdowns for each trade that can be aligned with work of other trades
- Follow building models (e.g., the best way to build an interstitial space)
- Establish flow patterns and rate of work

Guideline 3: Materials and Methods

Strategies that reduce field labor or reduce the cycle times of the trades are an effective method to streamline sequences and to align sequential activities. Recommendations:

- Eliminate fieldwork through prefabrication. Only assembly should be performed in the field. No shaping, cutting, or welding should be performed in the field if avoidable.
- Choose methods that ease the flow (e.g., rolling scaffolds).
- Perform work when it is easier.
- Plan material storage and movement.

Guideline 4: Reliable Work Assignments

Making work assignments for crews based on complete design information, available materials, and accessible workspace is a vital element of production. Sequence planning must endeavor to create a predictable work environment and establish reliable production rates for crews of sequential activities. Recommendations:

- Require formal planning at the foreman level
- Organize design information for use in the field
- Track reliability of work assignments

Guideline 5: Feedback

Data gathering on progress and potential disruptions allows continuous improvement of sequence plans. This feedback from the field plays a vital role in adjusting crew assignments and reducing interference problems that disrupt field operations. Recommendations:

- Use recurring update meetings
- Monitor all efforts of coordination, planning, and construction as well as the impacts the performance has on future work
- Identify hindrances through planning reliability checks and pass on the information to the general contractor

RULES TO SUPPORT SEQUENCING GUIDELINES

A detailed set of sequencing rules was developed as part of this study. By combining established sequence practices documented in the literature, feedback from contractors, and case study data, a structure for collecting and categorizing rules that support the five general sequencing guidelines was developed. The rules are categorized as follows:

- Design and detailing
- Coordination
- Planning and sequencing
- Materials and methods
- Crew management
- Information and feedback

Design and Detailing Rules

- The design team should involve all trades at the beginning of the project. This will allow all interdependencies to be raised and a solution developed that satisfies all the trades. As a trade is no longer affected by future decisions it is removed from the design team.
- The engineer should indicate the specifications for major equipment and allow the supplier/fabricator to develop the design and distribution systems to reduce the amount of duplicate effort.
- Mechanically intensive buildings should be designed structurally and architecturally to house mechanical systems. Vertical and horizontal distribution zones should be sized proportionally to the volume of systems needed in the building.
- The project design should utilize major components (assemblies) that can be preassembled to improve quality and reduce the amount of fieldwork necessary.
- Plenum spaces should be designed with adequate provisions for systems. The stratum of systems distributed in ceilings should be the same in all areas so trades can complete work in the same order in all areas.
- Details and dimensions should be standardized as much as possible. This will allow for maximum prefabrication that will reduce cycle times and improve productivity as well as quality.
- Design drawings should be produced as they are needed in the field so that the project is not overdesigned if the owner requires changes to be made. For example, if the floor-by-floor MEP drawings are completed before the foundation drawings, then the field forces will not have information to proceed. If the owner then wishes to issue a change, some of the time used to develop the MEP drawings would be wasted.
- The product and process designs should be completed simultaneously to improve performance. The crew's performance will be improved if

the method to construct a product or install materials is developed at the same time as defining the specifications of the final product.

- Design information should be formatted for use by foremen and crews. For example, piping designs are best communicated to foremen with plans and elevations for dimensional purposes combined with isometric drawings to clarify orientation. In contrast, electrical foremen require design information about circuit distribution, conduit routing, architectural plans, and wall sections to be organized by section.

Coordination Rules

- Architects and consulting engineers should provide CAD backgrounds that are drawn to scale, with common reference points for the coordination process.
- General contractors should manage the coordination effort and provide CAD drawings to mechanical contractors to turn over to plumbers and to electricians to add layers of detail. Mechanical contractors may have the responsibility to compile layers and plot drawings for weekly review meetings.
- Typical order of priority for design
  1. Drain, waste, and ventilation (DWV) may need to go first due to slope restrictions
  2. Sheet metal goes in next due to size restrictions
  3. Large domestic supply, conduit and cable tray, and sprinkler are in next as they can be rerouted more easily
  4. Smaller diameter pipe and conduit is routed last
- Ceiling should be constructed from the top down. Typical order of priority for construction
  1. Sprinklers often go first as they can be pushed close to the ceiling and in some cases need to be operational earlier
  2. Sheet metals goes in next due to size restrictions
  3. Large domestic supply, conduit, and cable tray are in next as they can be rerouted more easily
  4. Smaller diameter pipe and conduit is routed last
- Coordination drawings should include all items 4 inches or larger drawn to scale, and all other items drawn single line. Drawings should include accessibility buffers for maintenance and testing of systems.
- Common dimensional pitfalls
  1. Piping sizes do not match actual diameter:
     - ½ inch pipe = 0.84 inch outside diameter
     - 2 inch pipe = 2.375 inch outside diameter
     - 4 inch pipe = 4.5 inch outside diameter

2. Piping insulation, fittings, and supports
3. Ductwork insulation, flanges, and supports
4. Miscellaneous
   - Structural steel fire proofing
   - Code required clearances
   - Maintenance access clearances
   - Design basis equipment (size not known until late in process)
- Four types of skills are needed from each team member: (1) field experience to provide judgment on design detail; (2) decision authority for resolving changes; (3) CAD expertise for drafting layers; and (4) communication skills.
- Three-dimensional CAD modeling of building systems provides the most effective method to perform coordination, as it allows systems to be visualized and interferences to be detected between systems.
- On average, a building floor requires one week to produce CAD layers from each trade, one week to examine and resolve problems, and one week to redraw and correct problems. This three-week cycle will only expand with more complex systems, larger floor plans, and tighter horizontal and vertical distribution zones.

Planning and Sequencing Rules

- Once work is coordinated, a detailed work sequence should be developed in a partnering type of environment. The order in which specific work will be completed, work directions, and flow rates should be agreed on.
- Work should be coordinated and signed off before any construction is done in that area and should be completed at least two weeks before a contractor works in the area.
- An analysis should be performed to ensure that the work area is coordinated and ready and that materials and tools are on-site before the work is put on a weekly work plan.
- Materials should be prefabricated and, if possible, buried in slabs to minimize construction time, improve quality, and reduce coordination problems.
- Where there is no air conditioning, and therefore a significant part of MEP coordination issues are eliminated, there is less need to coordinate. Therefore there may be projects in which detailed coordination is unnecessary. However do not neglect planning of installation with other trades.
- Concrete slab placement and shoring removal often dictate the pace for the rest of the project. This activity should be used to establish a

steady release of work for layout and installation crews. Pouring large slab sections with less frequency can detract from flow while pouring small sections every day can increase flow.

Materials and Methods Rules for Crews

- Take measures to improve mobility of crews and materials. Examples include rolling carts, palletized materials that can be moved with jacks, and mobile bins for small tools.
- Choose materials fittings that reduce field labor.
- Use embedded anchors cast in slabs instead of anchors drilled or shot into slabs to save layout and installation time.
- Move distribution from walls to slabs and ceilings.
- Standard sequence goals
  1. Place as much material as possible on slabs.
  2. Complete overhead work prior to partitions.
  3. Phase partitions, building those with in-wall electricity and plumbing first, leaving those that block traffic for a second pass.
  4. Pre-load materials prior to completing walls.
  5. Keep floors clean. Debris restricts movement and can delay crews.
- Simplify the installation method as much as possible to minimize the amount of "looping" a crew must perform. Looping is the process of a crew having to set up in the same area multiple times to complete different steps of the activity. A simpler process would require only one setup. For example, the process of installing sprinkler pipe requires a crew to set up and install the main line, then come back and install the branches, and then back again to install the heads. If the crew could easily set up once in an area and install the main line, branches, and heads, the amount of time spent setting up would be minimized.

Materials and Methods Rules for Suppliers

- The "supplier" should consist of designer/fabricator/installer. This way there is no excess waste in the design, the fabricator creates a high-quality cost-effective product, and the installer has prior knowledge of the product when it comes to final installation.
- Feedback about work progress from on-site crews should be passed to fabricators. This will allow the fabricators to produce what is needed for the crew and not produce more than necessary. This will minimize waste if design changes are made, reduce damages to products, and minimize overhead due to reduced storage needs.

- Suppliers need to reduce their changeover times of fabricating product to reduce impacts of large batches. When a large inventory is created potential waste exists if a change is issued, and crews in the field suffer if one piece is missing and they have to wait until the next run to finish their work. For example, consider a mechanical contractor installing a run of ductwork. A fabricator could create all squared duct first, then all branch duct, and then all the turns. This would force the field forces to put up pieces of a duct run one portion at a time and expose the run to risk if there are field interferences and/or design changes. For best performance, the fabricator should be able to fabricate the run as needed in the field.
- General contractors should attempt to manage suppliers in tiers. The first tier should consist of a limited number of suppliers who in turn manage a larger set of second-tier suppliers. The first-tier suppliers will provide everything for the project and are directly coordinated by the general contractor.

Crew Management Rules

- The steps of a process should be synchronized so that, within the crew, one crew member is not waiting for others to finish.
- Collaborative work environments for management and crews should be developed. Make each crew's performance rating based not only on the performance of its crew but of the entire project. If one crew's ability to perform or to receive a satisfactory evaluation requires it to make work difficult for other crews, then there is no incentive to work in a collaborative manner.
- Feedback is required from site personnel on progress and problems in order to properly plan and make changes to the process.
- Multiskilled personnel should be developed to perform work in complex areas. This also aids in managing production by increasing flexibility to react to production needs and labor uncertainty (i.e., absences).
- Personnel should be developed so they can determine the origins of a particular problem. This allows for the problem to be tracked to its roots and then allows management to reduce the impact in the future.
- Projects should be planned for materials to arrive on-site "just-in-time" to minimize waste (e.g., financial costs, space, and damaged materials). Just-in-time is constrained by more than time of need, but more accurately described as materials being delivered when

economically practical, and when space in proximity to the point of need is available. All resources must be allocated in accordance with the reliability of their consumption.

- The cycle time of installing a product should be reduced or matched to related activities to allow for production to be better managed.
- The area of operations should be properly managed to keep a continuous flow of crews moving through the area until all tasks are completed. The flow should be sequenced to minimize interferences in installing products. The crews should be balanced to allow each crew to do comparable amounts of work in the same time.
- The lead crew should not get too far out in front as this will not allow the entire process to go faster, and may lead to waste and/or delays if the customer requires changes in the areas of completed work. The entire set of crews should not proceed faster then the production rate of the slowest crew. This may require resizing to balance crews. There should be a time buffer between crews to ensure that each crew does not run into the next and cause additional workspace interferences. This also accommodates minor variations in output and productivity.

Information and Feedback Rules

- A continuous improvement program should be implemented and include the crews, designers, suppliers, and other involved parties. Weekly planning should be performed formally and recorded.
- The second step of the weekly work plan should be to analyze results. The weekly schedule should identify if the activity was completed according to the plan. If not, modification that can be made to the process should be identified for future implementation.
- Foremen should identify why work could not be done. Where outside factors caused problems (hindrances), general contractors should be notified. Hindrance data from weekly planning sheets are highly valuable to general contractors, as they pinpoint problems and bottlenecks in the field that require attention and direction.

CONCLUSIONS

This chapter identified techniques to create productive sequences and the resulting benefits observed by project teams. The coordination process was found to provide the most effective means to shield sequence plans and the work of crews from disruptions. A set of sequencing guidelines were developed based on production management principles and the typical challenges of specialty construction. A detailed set of rules was assembled to drive production sequence development in the design, coordination, planning, method selection, and management of construction projects.

Contractors need to develop unique sequences for each project. While sequence models can help develop strategies for different types of buildings (e.g., interstitial labs, hotels, hospitals), a distinct effort must be made to develop a productive sequence after design and coordination efforts are completed. Pressures to perform "fast-track" projects in which construction begins before design is complete oftentimes mar this effort.

The greatest common difficulty found was the lack of cooperative planning during coordination and sequence planning followed by a lack of field directives from general contractors. This phenomenon has become more evident and problematic as a result of two visible trends: (1) specialty trades continue to become more specialized and (2) design consultants increasingly rely on specialty contractors to complete and coordinate designs.

The role of the owner in the sequence planning process could be elevated to play a more important contribution to the building process. One key area of impact for owners is the level of coordination assistance expected from design consultants. Another potential method for owners to influence the sequence planning process is to require formal coordination, sequence planning, and field directive from general contractors. Tools such as building information models (BIMs) can significantly assist contractors in the coordination process and owners can require all contractors to use them. Finally, owners should demand that horizontal and vertical distribution zones for mechanical and electrical systems be incorporated into architectural and structural designs with appropriate priority. Cost growth on projects with inadequate space for specialty systems can be expected. On the contrary, buildings designed with proper space for systems benefit from more productive work sequences, more flexibility for late changes and additions, and finally, lower building operations and maintenance costs.

5

IDEAL JOBSITE INVENTORY LEVELS

Philip E. Nimmo, *MCA Inc.*
Dr. Perry Daneshgari, *University of Michigan and Michigan State University*

INTRODUCTION

Inventory represents a significant investment of cash, labor, facilities, and equipment for all industries, including construction. The U.S. automotive industry has successfully reduced finished goods inventory from months to days. Leading retailers have successfully reduced inventory costs by eliminating large storage. Distribution centers and consumer goods manufacturers have achieved inventory levels so low that they are nearly virtual. The common element in all of these improvements is that the true purpose of inventory has been recognized and that they called on the entire industry supply chain to help achieve the needed results. Contractors handle large amounts of expensive materials and have generally paid little attention to the total cost of this practice within the industry.

Inventory is nothing more than a small stockpile of the resources needed for the system to function; in this case, the materials needed for the craftsmen to continue working productively. When viewed as a buffer, or a tolerance, inventory really serves to accommodate acceptable variation in demand. Much like the size of the gas tank on an automobile determining the distance between fuel stops, the jobsite inventory determines the time between potential work stoppage. Logically, foremen believe that the bigger the inventory buffer, the less likely a work stoppage is to happen that day. In reality, the size of the jobsite inventory has little to do with efficient job flow. Just because a small truck with a large gas tank is capable of driving more than 1000 miles between stops, this does not necessarily mean that the truck is fuel-efficient. Increasing endurance does not guarantee fuel efficiency, nor does increasing jobsite inventory guarantee productivity or material availability. By increasing jobsite inventory we do not decrease the risk of productivity loss, we simply shift the risk away from too little material. Risk of work stoppage due to inadequate quantity of material is replaced by increased risk of work stoppage due to incorrect material and excessive material handling. This practice depletes cash resources from the contractor and throughout the supply chain. By increasing jobsite material inventory we reduce the financial inventory to compensate. Additionally, by increasing the jobsite inventory we increase the labor inventory needed to handle material. An extension of this observation is that the investment of labor in jobsite inventory increases with increasing amounts and lengths of time that inventory is located at the jobsite. Material that is not installed when it is received needs to be organized, stored, managed, and moved. Each time one of those steps is performed, an investment of labor time is added to the total cost of the inventory.

INVENTORY PRINCIPLES

Traditional efforts to reduce the cost of having large jobsite inventory have been aimed at shifting the burden from the contractor to the distributor. Such concepts as consigned inventory and same-day delivery have accomplished little more than shifting the expense from the contractor to the distributor. The distributor is therefore forced to increase warehouse inventory levels to accommodate the demand for reduced delivery times, and still satisfy the installer's need in spite of the variation of daily work changes on the jobsite. At the distributor level the inventory increases must be of sufficient magnitude to accommodate the variation associated with all of the projects of all of the customers at any given time. This equates to a large buffer and reduced accuracy with respect to what items and quantities should be held in that buffer, representing a stacking of tolerances phenomena within the logistics and material management business. The jobsite has a buffer to accommodate variation in work performed, the contractor's warehouse maintains an inventory buffer to accommodate variation in the delivery performance and responsiveness of the vendor, the distributor maintains an inventory buffer to accommodate the collective variation in planning and communications between themselves and all of their customers, and finally the manufacturer maintains a small inventory to allow for steady flow and provide a competitive buffer against sudden fluctuations in demand. Figure 5.1 shows the flow of material from the manufacturer to the contractor's jobsite, using all of the buffers identified. The flow is cumbersome, and is often circumnavigated using expediting processes and third party shipping, all of which add considerable and often hidden costs to procurement.

Historically, contractors have allowed their field supervisors and project managers to determine the necessary material types and quantities required for their projects to be purchased without providing specific material management or procurement logistics training and education to these employees. Furthermore, supervisors and managers do not receive training in simple statistical methods that would help them recognize and reduce variation. The presence of variation in most of their work is predictable. The sources of variation are separated in both time and location from the symptoms, making it more difficult to identify the actual causes. Therefore, the variation that causes the need for increased material inventories is not recognized—such as variation in order lead times, change orders, schedule changes, labor quality, initial estimates, and the need to handle material inventories. The result is that inventory levels within the construction industry have remained high and costly. Prior research has shown that approximately 40 percent of a craftsman's time is spent handling material. Without adequate training, the field supervisor maintains an inventory level that provides a personal degree of comfort. Often, this is based on previous experience with material shortages from underestimation of the actual consumption rate. In fact, it is common practice for a contractor to purchase and ship as much as 80 percent of the anticipated material to a jobsite at the time of initial mobilization. By reducing the material inventory on the jobsite, a contractor can recover some of the 40 percent of labor that is spent handling material, reduce the risk associated with damage or injury, and reduce their cash conversion time, while still maintaining adequate inventory levels to avoid starving the labor.

We know from previous research that a craftsman's time spent handling materials is roughly 40 percent of the total hours worked. We also know that some of the major contributors to nonproductive time are related to a lack of available material at the installation location. In an effort to alleviate the nonproductive time, most foremen and superintendents have adopted a practice of using jobsite

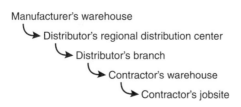

Figure 5.1 Supply chain flowchart

inventory to form a buffer against shortages of material. Inventory should form a buffer; however, due to the difficulty associated with prediction of work and frequent changes, as well as the perceived slow response time of the supply chain when adapting to these changes, many field supervisors have steadily increased their on-site buffers to a point where the 40 percent figure becomes very real and very common. Labor is now being overburdened with the costs associated with storage and handling of these on-site material buffers.

The goal of this chapter is to examine the principles used within and outside the construction industry, which can allow us to reduce the jobsite inventory levels and still maintain an adequate material availability at the point of installation. To identify the ideal inventory level, this chapter concentrates on:

1. Understanding the current construction industry's procurement and material management practices
2. Using the practices associated with principles applied in other industries, including manufacturing, retail, automotive, or information technology to better understand what the proven potential benefits from reductions of inventory are, so that contractors may achieve them
3. Reducing business risk by identifying practices that encourage the reduction of inventory and that result in cost savings and increased profitability for the contractor
4. Reducing waste by using effective and efficient material logistics planning and management techniques
   - Visibility of the procurement process and function through use of a standardized process of project management, and trend monitoring for decision-making based on statistical process control (SPC) techniques
   - "Gemba Kaizen" improvement from the customer's perspective; methods for early awareness and communication of changes and customer's intent through accurate and timely short interval planning and three-day look-ahead schedule
   - Use of buffering inventory levels that are based on application of queuing theory and an understanding of the loss function and economic order quantity

This chapter provides a recommendation for the ideal jobsite inventory level and suggests a methodology for achieving and maintaining operations at this level. Our investigation was conducted by the use of an extensive literature review, two industry-wide surveys, more than 50 jobsite visits with supervisors at more than 20 geographically distributed electrical contractors, and interviews and observations of practices in other industries.

The results of our research indicate that there exists both a theoretical and a realistic ideal inventory level for jobsite materials. Both follow the principles described by the loss function shown in Figure 5.2. The loss function shows that there is an ideal amount of inventory to maintain on the jobsite. When the amount of inventory is either less or more than the ideal, the contractor incurs unnecessary costs. As inventory levels go above the ideal there are increasing costs resulting from factors such as increased storage, extra handling, additional breakage, and mismanagement of cash that is tied up in inventory, as well as a greater susceptibility to hidden costs from change orders. In the same way, inventory levels that are lower than the ideal also increase cost. The low-inventory costs include lost labor time, missed schedules, increasing damages and claims, and diminished reputation. The further away from the ideal inventory level in either direction, the greater the impact of these factors and the higher the costs.

Principles such as queuing and the customer point of entry will be the driving factors in achieving improvements in the total system performance. The benefits of these performance improvements will be realized through increased first-time pass (FTP) (error-free, on-time delivery of the correct material to the correct location at the correct time), and increased productivity through waste reduction (eliminating the unnecessary and redundant valueless activities); all visible in the form of reduced jobsite inventory levels. Perhaps equally significant are the performance and financial benefits that result from the reduced inventory, such as increased cash flow, decreased cash conversion time, reductions in returns and damaged material issue resolution events, and improved safety and injury cost.

The results of our research support and validate the hypothesis that there is an ideal level of jobsite inventory. The level is best expressed in terms of days and

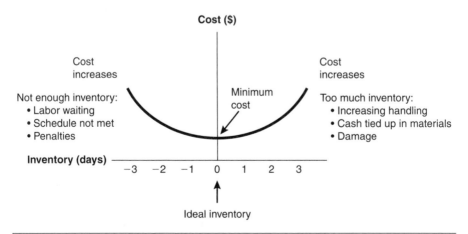

Figure 5.2 Inventory loss function

when viewed in this manner tends to vary little with respect to the number of men, type of material, and stage of the job. Factors that influence the ideal number of days of inventory on the jobsite include order fulfillment time and rate from the distributor, accuracy of pre-planning, and coordination efforts with the general contractor and other trades. A perfect procurement process that reliably delivers any and all material exactly when and where it is needed would allow the on-site inventory to be reduced to a theoretically calculable value. By balancing the common, economic order quantity equations with simplified queuing theory equations, we could arrive at such a total. However, given the number of unpredictable variables that exist in construction, we will settle for a fundamentally sound target range as opposed to a specific ideal value.

For the purposes of this chapter, the term *ideal* is defined as the nominal level of material that is sufficient to prevent an excessive risk of exhaustion without creating an undue expense. The impact of these various cost drivers within the industry can and will appear anywhere that inventory is accumulated. During our initial investigation it was revealed that there are numerous inventories within the construction supply chain. The relatively common inventory locations encountered include:

- Manufacturers
  - Raw material inventory
  - Work in process
  - Finished goods inventory
- Distributors
  - Regional distribution centers or zone warehouses
  - Branch warehouses and retail store shelves
  - Consigned inventory at original equipment manufacturer (OEM) and large project sites
- Contractors
  - Shop/warehouse inventory
  - Jobsite inventory
    - Trailers
    - Lay-down areas
    - Service vehicles
- Customers
  - Spare parts stock

Although this may initially appear to be an excessive list of inventory locations, the efficiency is not necessarily tied to how many buffers exist; rather it is related to how efficiently they are managed. As an example, we can look at a similar process, flow of water to a community. The source of fresh water for many communities is accumulation of precipitation in higher elevations. Snowfall on a mountaintop may accumulate for several months during the winter creating a large buffer of available fresh water; this is similar to a manufacturer who tools for, and produces parts in a batch process. When an order is placed for material from a manufacturer, this represents a simple catalyst that initiates the flow of material; this is the same as a spring thaw beginning the flow of water runoff from our mountaintop.

The water flowing down our mountain will, of course, follow the path of least resistance, the path that removes the most available water the fastest from the top of the mountain without regard for where it is going to end up. Similarly, when manufacturers receive orders they are anxious to ship material by the most expeditious means available; often third party carriers are ready and waiting. Because the residents of our community really do not want to open their faucet in the spring and have a river flow into their home, or open their faucet in the dry heat of summer and have no water available at all, the community erects a dam to form a reservoir. The reservoir is simply a buffer to allow the flow of water to the community to be managed and metered in accordance with the demand. At the reservoir level, water is shared between multiple communities similar to our distributor who maintains an inventory to buffer the demand against the batch manufacturing processes. Contractors routinely maintain small inventories near the point of installation to compensate for short-term changes in the demand, shifting schedules, change orders, or variation in productivity. The contractor's small local inventories are similar to a community water tower that is fed from the reservoir and used to accommodate small fluctuations in demand at the individual user level. Finally, we mentioned that customers often retain a small stockpile of spare parts for their spontaneous maintenance needs; this is exactly the same as a resident keeping a pitcher or two of fresh water in the refrigerator where it is readily available exactly as desired.

The point in this comparison is to emphasize that construction is not suffering from an unprecedented number of buffers. Therefore, improvement in profitability will not likely come from elimination of the inventory locations; rather it will come from improvement in efficiently managing the supply chain. The factors that will lead to sustainable improvement are related to the application of such fundamental principles and concepts as:

- Waste reduction (effectiveness of planning and ordering; jobsite layout and procurement logistics)
- FTP (quality of the installation)
- Labor productivity planning and management (variation in material consumption rate)
- Project productivity tracking and management (variation in the number of craftsmen; combined material consumption rate)

- Change orders, customer point of entry (variation in scope of work)
- System cycle time (supply chain responsiveness)

DEVELOPING THE LOSS FUNCTION

Left Side

On the left side of the loss function diagram (Figure 5.2), we suggested that too little material inventory increases labor cost to the contractor. Although this is a well-known fact, efforts to quantify the cost of not having material have been somewhat of a matter of circumstance. If we accept that the foremen, in nearly every case, are actually planning the next day's work with primary, alternate, contingent, and emergency plans (PACE), then the cost is rarely that no work gets completed for the day. However, the cost associated with shifting workers and material between these plans can be approximated.

To quantify this cost let's look at a simple example. Assume that a particular jobsite is running with ten craftsmen. The jobsite contains an inventory consisting of most of the anticipated commodity items (the project manager ordered 80 percent of the material identified on the estimate at the start of the job; however, the estimate was not broken down to a fine detail); therefore, each day a small shipment of miscellaneous items is required to keep things moving effectively. Additionally, let's assume that most, but not all, of our deliveries are received as expected and as ordered, and that our loaded cost for labor is $40 per hour.

If inconsistencies and last minute changes to the order receipt at the jobsite cause the interruption and realignment of workers resulting in a loss of productive effort equal to one hour per week per worker, then we can use this as a basis for calculation. One hour per week per worker is simply allowing twelve minutes per worker each day to finalize whether he or she will be working on the primary plan, alternate plan, contingency plan, or emergency plan for that day, and allowing them to gather the correct material to get started. Our research indicates that this is a common occurrence throughout the industry.

One hour per week per worker, for our ten-worker crew at $40 per hour equates to:

$$\text{Loss Productivity} = (1 \text{ hr/worker})(40\$/\text{hr})(10 \text{ workers})(50 \text{ weeks/year})$$
$$= \$20,000 \text{ per year} \quad (5.1)$$

This same example can be expanded to measure the entire company. If the company employs 200 craftsmen at $40 per hour, then the resulting cost of lost productivity is $400,000 per year. Again, this example is only illustrating the cost associated with the current practice of "PACE" planning, combined with daily determination of which plan is going to be implemented.

Having established that there is a cost associated with the left side of the loss function, too little material (or lack of the correct material) on the jobsite, we still have to establish the shape of the curve on the left side. As we move farther from the ideal, toward too little inventory, how does the cost increase? To answer this question we need to understand the inventory locations in the supply chain before the jobsite. In many cases there are material buffers at the contractor's shop, at the local distributor branch, perhaps at a distributor's regional distribution center, and finally at the manufacturer.

Generally speaking, orders placed for same-day or next-day delivery will have to be filled from the contractor's shop or the distributor's local branch. Orders placed for items that are not in stock (either items ordered or quantities ordered exceeding availability) will have to be filled from the distributor's regional distribution center (if they have this type of structure). Filling orders from a distribution center generally requires two days' notice (one day to the branch and next day to the jobsite). Any order placed for material that cannot be filled by the distribution center (or from a distributor that does not use a distribution center structure) will have to be filled from the factory's finished goods inventory. This scenario invariably involves third party carriers and a minimum of three days to arrive at the jobsite.

With respect to the shape of the loss function curve, it is driven by the likelihood of encountering unavailable material at the jobsite. Unavailability is going to reflect the inverse of availability. If we work from left to right on the loss function, if material is not in stock at the jobsite on the day that we need it, then we are guaranteed not to have the material at the jobsite; the cost is infinite. Most typical daily orders are filled from branch stock, so there is a rapid reduction in the unavailability of material as we move right. The branches generally contain about one-tenth of the material that the distributor's distribution centers maintain. This results in the flattening of the curve as we pass through the point equal to two days' advance ordering. However, the material is still reducing the unavailability as we maintain more material on the jobsite and order less frequently and in advance. Figure 5.3 illustrates the left side of the loss function as we have described it here.

Right Side

Studies of distributors reveal that as much as 7 percent (average of 2.5 percent) of the material that they sell to contractors is returned. Many of the most common reasons for returns include:

- Incorrect material was ordered
- Incorrect material was shipped
- Changes to the plan require different material

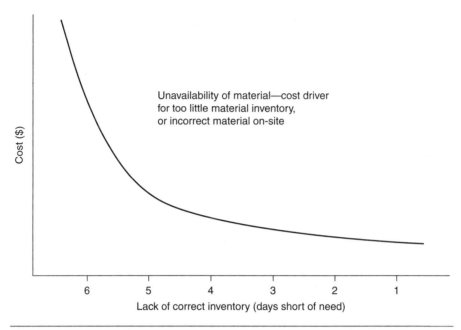

Figure 5.3 Developing the left side of the loss function

- Material was not needed, too much material was maintained for comfort
- Inadequate space to store the material
- Delivery time was not when specified
- No way to handle the material as shipped

Building the right side of the loss function curve is completed similarly to developing the left side. In this case the obvious driver for the extreme right of the graph (too much inventory) is returns, material that was on the jobsite and never needed to be there at all. Returns represent a unique case of the cost associated with material movement and handling. When the return occurs there is no longer a justification for the labor spent to handle and move the material at least twice. The calculation for this labor cost follows the same logic that was outlined previously while constructing the left side of the loss function curve for labor waiting on material. At this time we will look at the extreme case where returns are involved to identify the cost beyond the labor.

To identify the cost to the contractor we need to begin by understanding the cost to the distributor. Returns, from the distributor's perspective, are the worst form of rework. They represent material that was purchased, ordered, and received into inventory, stocked on the shelf, sold, and picked, packaged, delivered, picked up, re-received into inventory, and restocked on the shelves. Contractors often expect this service with full credit for the items returned. As organizations strive to implement six-sigma initiatives to improve the quality of their operations, this form of rework is the worst hindrance to a distributor. A high-quality transaction for a vendor is going to achieve error-free delivery of the material required at the jobsite. The FTP rate associated with distributor operations rarely exceeds one or two sigma (~95 percent FTP). The limiting factor for the distributor is often related to their external interface with the contractor, more so than with their internal capabilities. A 3 percent return rate will invariably guarantee that no distributor will ever achieve three-sigma (~99 percent FTP) quality performance.

Figure 5.4 identifies the flow of material through the process described previously, and an approximate cost to the distributor for the labor associated with each step. Notice that there are 7 steps required fulfilling the incorrect order, plus 11 steps required returning the incorrect order and fulfilling the correct order. In addition, while the distributor is correcting the error, it cannot use those resources elsewhere. Therefore, there is an opportunity cost associated with these 11 steps as well. Therefore, the total cost for the distributor includes 29 steps versus 7 steps if the order had been correct in the first place. Assuming a cost of $10 per step, the additional 22 steps represent an extra cost of $220 to the distributor.

The distributor has spent $220 extra to satisfy the contractor, who now wants 100 percent credit for the returned items. Although the credit appears, the cost seems to vanish; where does it go? Obviously, this cost is inherently built into the selling price of every item that the contractor purchases. For a simple example, let's assume that the contractor returns 3 percent of the material purchased on each job. Using the data from the illustration, this says that 97 percent of the orders are filled at a cost of $70, and that 3 percent of the orders are filled at a cost of $290. The distributor's cost of processing orders in this fashion is:

$$\text{Actual Order Cost} = (0.97)(\$70) + (0.03)(\$290) = \$76.60 \quad (5.2)$$

This indicates that the distributor is paying $76.60 for processing orders that could cost only $70 to process. Logically, for distributors to remain in business and to profit they must increase the selling price to account for this increase in cost.

With respect to the shape of our loss function curve, we can also build the right half from left to right (Figure 5.5). On the extreme left there is no material on the jobsite and therefore there is logically no labor cost associated with having too much material. As we move right, increasing inventory on the jobsite, the cost begins to climb. The rise is due predominantly to increased material handling and movement expense (this can be approximated using the methodology shown for the left side of the loss function), and also due to the increasing likelihood and subsequent cost of returns. The shape will therefore be similar to the left side;

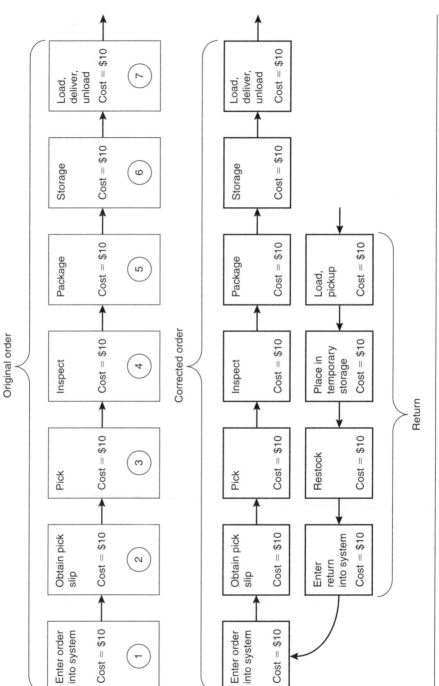

Figure 5.4 Cost to the distributor of rework due to returns and refills of orders

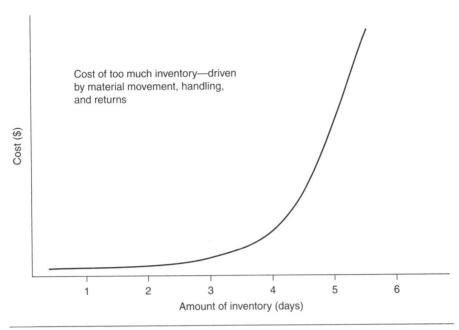

Figure 5.5 Developing the right side of the loss function

however, it will rise somewhat sharper as return costs add to the contractor's labor cost.

At this point we have constructed the loss function curve in terms of defining the limits of both the left and right sides of the curve, and we have also established that the left side has a relationship clearly defined between each day of reduced inventory and the rising cost. Based on the availability of materials from the branch, distribution center, and manufacturer, the likelihood of unavailability given two days is approximately 8 to 12 times (depending on the size of the branch) that of next-day planning, and given three days most distributors can fill any order for any item except long-lead-time or special builds. The right side is a little more difficult. However, the research does allow us to quantify the incremental cost increase associated with too much inventory. Based on data from look-ahead processes, and verification with representatives of numerous contractors, we have established that by using the PACE method foremen are able to reliably achieve approximately 80 percent accuracy on a three-day forecast of scheduled work to be performed. Next-day accuracy increases to roughly 90 percent. The foremen have also indicated that the accuracy of a scheduled work projection five or more days in advance is almost certain to have some degree of change. If we fit the shape of the curve described to these points, then the complete loss function curve can be drawn, with a minimum total cost near the point of three days of inventory.

However, this requires that the three days of inventory coincide with a three-day planning and replenishment strategy.

Inventory Management Principles and Practices in Other Industries

With our development of the loss function for jobsite inventory pointing us toward the ideal operating point, we need to begin looking at how to get there. The first step is to see what has been utilized in other industries to move in this direction. Our study of other industries to identify practices and benefits from material inventory reduction and ideal levels included analyses of several major and recognized leaders. We investigated supplier relationships and procurement practices of Wal-Mart, the manufacturing and material work in progress (WIP) practices at Dell, and the finished goods inventory levels and practices of the big three automobile manufacturers. Wal-Mart has built a procurement system where the product manufacturer owns, and in many cases handles, the inventory on their shelves. Additionally, Wal-Mart was a pioneer in the application of a practice known as cross-docking, where distribution centers are not used to establish independent material buffers, but only to facilitate the prepackaging or kitting of materials needed at each specific location, regardless of their point of origin.

Dell Computer has focused on minimum on-site inventory to reduce the cost of their WIP. By lowering the installation site inventory from a historical level of 35 to 40 days down to a more cost-effective level of less than 5 days, Dell has benefited from reduced handling costs, reduced risk of obsolescence, and reductions in the use of critical financial resources, while improving the system response time to customer orders. Dell's practices allow the environment to change frequently and drastically without creating undue cost or risk to the company. The focus on robust procurement and responsive material management has allowed them to become a low-cost, high-profit producer in their industry.

Automobile manufacturers have focused heavily on the value of finished goods inventory, recognizing the significant investment in both material and labor that is being stored and handled prior to sale. In the case of a contractor, this is similar to maintaining large quantities of material at the jobsite either long before it is needed, or long after the need no longer exists. The automobile manufacturers have successfully reduced their inventory levels resulting in improved cash flow and improved responsiveness to changes in customer demand.

These three examples provide powerful illustrations on the benefits associated with the application of fundamental business principles in the practice of inventory management. In the Wal-Mart example, the emphasis is on waste reduction. An unnecessary buffer is reduced or even eliminated in an effort to save money and improve overall system performance. In the case of contracting, the reduction of distribution warehouse inventory could be equated to two low-value activities: (1) the storing of material at the contractor's warehouse and (2) the maintaining of large stockpiles at the distributor. In an ideal model, neither of these warehouses would serve to duplicate the buffer of material needed to accommodate variation in demand, but would more effectively be used to add value by simply assembling the materials from various manufacturers and packaging them in shipments of kitted material with deliveries timed to correspond to the actual installation demand.

OPTIMAL INVENTORY LEVELS IN CONSTRUCTION

Achieving optimal inventory levels in construction must be accomplished with today's supply chain. However, we also cannot assume that the supply chain's current practices are rigid and incapable of improvement. The minimal pre-planning of material needs, independent of company policy, has created an environment that is prone to back orders, unavailable materials, long lead times, and third party direct shipments. These are all known to reduce the effectiveness and productivity of construction workers, and increase the cost of material handling to the contractor. At this time, interviews with distributors and suppliers have indicated that most stock items, regardless of inventory level, can be acquired for delivery to the jobsite with three days' advanced planning.

Given that the fulfillment rate nears 100 percent at three days, and that the cost of delivery is typically insignificant in relation to the cost of handling material on the jobsite, and the lack of current predictability of work to be performed with any reasonable precision is not currently achievable with forecasting beyond a few days, then the ideal level of inventory for the current procurement system is roughly three days. However, given that the current procurement system is not unalterable, the ideal level for jobsite inventory could not only theoretically, but realistically, be reduced to much less than three days. Achieving results in this fashion would require application of the principles described earlier that have been proven effective in other industries, and which are gaining popularity within the construction industry. The operational model for achieving these results will require a procurement process covering the following elements and practices:

- Planning
- Scheduling
- Delivery
- On-site material movement
- Returns
- Lessons learned and postmortem

Visibility of the Procurement Process

Visibility of the process and function through use of a standardized process of project management and trend monitoring for decision-making based on SPC techniques are key elements to improve material inventory levels. To properly implement systems and processes they must be visible to the operators, such as the project managers and foremen. Figure 5.6 is a simple trend monitoring run chart which shows the effectiveness of company policy to reduce the number of deliveries from vendors and to provide advance notice to vendors for all material orders. In this case, it was stated that the goal was to achieve one or two deliveries per week to any jobsite, and the enticement was that given three days' notice there will be zero back orders on stock items, and increased opportunities for kitting, packaging, and staged deliveries. The chart illustrates a partial success. The total orders decreased significantly as can be seen on the dark line. Over the three-month period the total orders decreased from 120 orders per week to just over 30. Unfortunately, the often divergent saw-tooth pattern shown on the next-day and second-day orders indicate that the users were not able to see the benefits of advance notice, and consequently struggled to achieve the goal. In this example cost savings were realized by both the contractor and the vendors due to waste reduction. However, benefits of increased efficiency from process improvement were not achieved during this sample period. This illustration shows both the benefits of proper visibility, as well as the power of analysis using simple SPC concepts.

Gemba Kaizen

Gemba Kaizen is a Japanese term that quite simply refers to improvement from the customer's perspective. Methods for early awareness and communication of

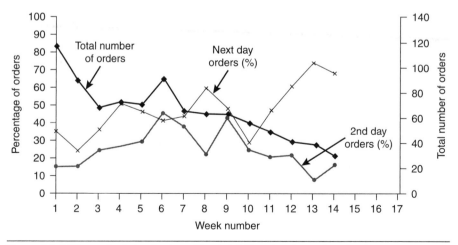

Figure 5.6 SPC run chart showing benefits of visibility related to reduced orders

changes and customer's intent are a critical element when it comes to maintaining the proper amount of the correct items on the jobsite. Through accurate and timely short interval planning and three-day look ahead, the foreman has the opportunity to relay information regarding jobsite changes and issues to the company and to the vendors.

Ensuring that the craftsman has what is needed, when and where it is needed, packed and labeled properly, and good quality, requires that all the elements of the system required to achieve this are informed about what the craftsman is going to do. Achieving the level of predictability required to reduce material inventory to their ideal levels necessitates an understanding of the factors that cause variation and degradation of the accuracy of the short-term plan. One such representation of these factors is shown in the Pareto chart in Figure 5.7. This chart quantifies the root cause for workers not being able to perform their work as originally planned. The importance of this information is critical when it comes to designing the operating, communications, logistics, and procurement systems to meet the needs of the installer. Equally important is an understanding of the relative impact of these inaccuracies on the profitability of the contractor. Figure 5.8 begins to translate the areas in which the unpredictability of work caused changes in actual versus scheduled work, impacting the number of labor hours spent working at less than optimal efficiency.

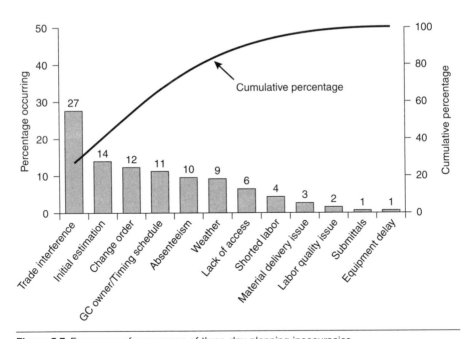

Figure 5.7 Frequency of occurrence of three-day planning inaccuracies

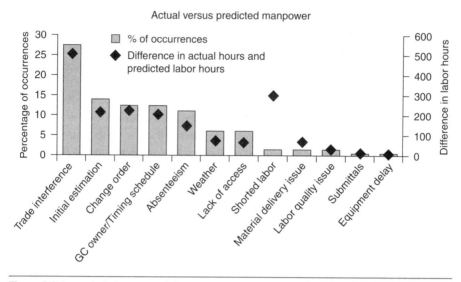

Figure 5.8 Impact of changes on labor

OPTIMAL MATERIAL MANAGEMENT MODELS FOR CONTRACTORS

The potential benefit for the contractor and the contractor's material management partners can be shown graphically by looking at the labor-planning diagram. Combining the historical or traditional S-curve for labor loading with the material planning and delivery results from this research, we can identify and begin to quantify a significant opportunity for savings associated with jobsite inventory reduction. Using a representative example, a contractor who orders and receives 80 percent of the material required for a project at start-up and then orders additional material in proportion to the manpower throughout the life of the project will purchase material as depicted by the inventory purchased line in Figure 5.9. Notice that the line for inventory purchased exceeds the 100 percent mark by approximately 3 percent; this is to reflect the material that will be returned. No allowance has been given for scrap, theft, loss, or other disposition of material; in this illustration it is either installed or returned. The inventory used line represents the cumulative installed material; that is assuming that material is installed in proportion to the S-shaped labor loading. The area between the two lines represents the uninstalled material on the jobsite. The area between the two lines is approximately equal to 47 percent of the area under the total inventory purchased curve. Based on this overlay, the potential for savings could be as high as 47 percent of the current cost of handling, storing, and returning material at the jobsite.

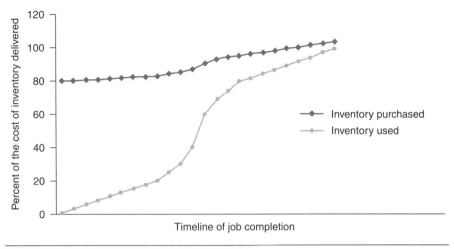

Figure 5.9 Jobsite inventory procurement

There are several factors that relate to establishing the optimal material management model, such that the ideal jobsite inventory becomes practical, achievable, and desirable. Improved construction system operations from application of several fundamental business principles will be necessary to structure the contractor for successful inventory reduction. The key principles are:

1. Waste reduction-using vendor partnership and improved communications with suppliers and manufacturers to reduce unnecessary material procurement costs
2. First-time pass-using planning, tracking, and forecasting techniques to ensure that the work being performed on the jobsite is being performed according to a plan and free from errors or valueless productivity variances
3. Customer point of entry-ensuring that the contractor is able to accurately anticipate the work to be performed to satisfy the customer's demand at any point in time, and to have the resources available to successfully satisfy that demand

SIX STEPS OF THE PROCESS OF PROCUREMENT

Accomplishing the ideal inventory through implementation of the three principles identified previously will be performed differently by each contractor. The company owners and leaders will create a specific culture that is based on their personal philosophies and ideals; these will always be visible in the resulting operating processes and policies of the organization. Accepting this variation in the actual models, the application of the principles will follow a set pattern to achieve the desired results. The remainder of this chapter will focus on the common elements of an implementable procurement process that enables operation within the ideal range of jobsite inventory, without undue risk or expense.

The procurement process can be broken down into six primary steps or areas of focus for the design of the business system. These six steps are illustrated in Figure 5.10.

Step 1: Procurement Plan

The purpose of the procurement plan is to identify the material items and project or jobsite characteristics that will contribute to the difficulty of the project. In this case, difficulty would be anything that represents an increase in risk. Greater likelihood of missing deadlines, not achieving the entire scope, not satisfying the target value to the customer, or failing to achieve the maximum available profitability would all constitute an increase in risk. The purpose of the procurement plan is to identify the risk drivers from the perspective of each stakeholder. There are three primary categories of risk associated with material procurement:

1. Business risk-associated with material pricing or price fluctuation, the cost of vendor services versus handling and storage at the contractor's site, and the cost of warranty and contractor commitment to service issues
2. Technical risk-associated with the accuracy of the take-off, the completeness of the system and its design, the correctness of the material, the proper function of the installed components, and compliance with code, specification, and submittal
3. Integration risk-associated with the movement of material, installation methods and techniques, productivity and trade coordination, communication on the jobsite, proper tracking of materials, maintenance of the inventories, and jobsite safety

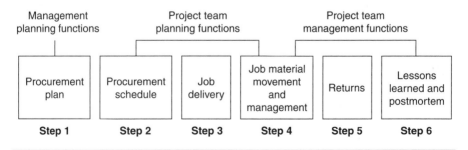

Figure 5.10 Six steps of the process of procurement

It is not sufficient for each stakeholder to identify their own risk drivers independently, since the interfaces between the organizations will impact the overall risk, both positively and negatively. By working together in the planning phase the contractor will have an opportunity to communicate the concerns of the installer, and collaborate with the manufacturers, manufacturers' representatives, distributors, and the general contractor or owner to resolve the technical and integration issues. Effective solutions nearly always require the cooperation of all stakeholders; anyone working for their unique benefit at the expense of the project will negatively impact everyone's performance.

To correctly implement a procurement planning process, the contractor must establish and communicate a few simple elements in a manner that is appropriate for their business as well as common and predictable throughout their business. These required elements are:

- Method for selecting the vendor(s) for the project
- Identification of who, within each stakeholder's organization, will be involved in the planning process
- Agenda and structure for a procurement planning event
- Method of communicating and achieving consensus on critical issues, including
  - Material list—source(s)
  - Work breakdown structure for the procurement plan
  - Jobsite layout, constraints
  - Special handling or delivery requirements
  - Submittal management plan
  - "Who does what"—scope of work for procurement
  - Material ordering and requisitioning structure
  - Billing and payment terms

The contractor must also develop an expectation with regard to when these tasks will be completed. The timing could be relative to the bid date, award date, or the project start date but should be consistent and predictable. The timing of the plan should also accommodate the needs of each stakeholder, in terms of time to perform their tasks to expectation of the system. This step requires the involvement of all stakeholders, and should therefore be designed to include the needs of the other stakeholders, not only the specialty contractor.

Step 2: Procurement Schedule

In the same fashion that the procurement plan has been designed to coordinate the needs and the scope of work with regard to material procurement and movement, the schedule forms the basis for coordination of the accomplishment of this plan. All stakeholders will play a part in maintaining and achieving the schedule. The procurement schedule is not the responsibility of the contractor or the vendor; rather it is the responsibility of the project team. The project procurement schedule should be as detailed as possible. The procurement schedule is developed beginning with the contractor's project schedule and works to ensure that material is available at the time needed.

For nearly all contractors this will be the most difficult part of achieving an ideal jobsite inventory. There exists a common awareness that project schedules are always changing, and often without regard to the trade contractor's concerns. For this reason many trade contractors tend to skip the scheduling piece all together, or at least leave out the pertinent detail until it is too late for the supply chain to react. To take advantage of vendor services and to reduce inventory without risk of starvation, it is imperative that the material be scheduled. At any time the schedule will only be as accurate as the available information; however, when the process is designed to allow communication of plans and changes then the impact can be minimized.

The schedule for procurement should expand the "who does what" portion of the plan to include "when." This simply says that to achieve on-time delivery of specific items, to specific locations on the jobsite, packaged, kitted, and labeled to meet specific expectations, the timeline should be shared with all stakeholders. There are no guarantees; however, it is logical that if the general contractor is aware of the lead times for not only gear and fixture packages, but for all specialty contractors' items in each area of the project, then they will be less likely to make unrealistic changes. Much of the reason that we experience the changes is because we have historically accommodated them, at the trade contractor's expense.

Recall that we are describing the development of a procurement process, not a specific project planning session. To benefit from this process it needs to become the way that the organization operates, not the way in which one project is managed. If only an occasional project attempts to use this process and these steps, then the entire procurement process that we are describing is an exception, and this adds cost to all stakeholders. For the system to be effective at reducing cost, it must be designed in a fashion that roughly 80 percent of the company's projects will use this process. Let the other 20 percent be the exceptions. A commitment to create, communicate, and follow this schedule has to be company practice, not just a written policy statement.

Step 3: Job Delivery

The procurement plan focused on identification of the risk, the procurement schedule focused on the timeline, and the job delivery step needs to focus on "where and how" material is to be delivered to the jobsite. This is the most critical step for achieving installation efficiency, first-time pass improvements, and productivity benefits from reduced inventory. Part of the project team's planning function should include a discussion with the foreman regarding material delivery. The foreman needs to represent the installer of the material; he is the point of installation and essentially the customer of the procurement process. The foreman should be the final authority with regard to requisition and delivery dates.

The job delivery step of the procurement process has to address several key elements of material preparation and movement. The main issues to be included in the design of this portion of the process are:

- Initial delivery dates and material list
- Means of updating the planned delivery dates
- Procedures for requisition or release of materials
- Procedures for changes to the material list and project scope
- Method of communicating the status of jobsite layout and changes to the layout
- Detailed procedures for communicating kitting, labeling and prefabrication expectations
- Detailed procedures for communicating and handling back orders
- Constraints on delivery, time of day, type of vehicle or specific driver
- A procedure for monitoring the effectiveness of the deliveries, and the interface between the stakeholders at the jobsite

This step has such great impact on productivity and in turn on profitability that there is a strong need for unique thinking. The ability to develop a process that allows an opportunity for the foremen to be involved with the delivery planning activity prior to the start of the work has proven to have significant benefits to the contractor.

Step 4: Job Material Movement and Management

This is the step where we have to account for several political factors and contingencies. There is no "right" or "wrong" answer. The procurement process has to acknowledge that union contractors face different agreements and contracts than nonunion contractors. Even within union contractors, there are different agreements in different geographic areas and for different types of work. In some cases and in some places the vendors will be able to place material directly at the point of installation; in other cases this will not be so easily accommodated. The company-wide process has to be agile enough to respond to changing circumstances. At the least, this step is an opportunity to plan and communicate contingencies for conditions that prevent the ideal delivery of material to the jobsite.

In cases in which it is possible to use vendors or other material handling experts to move material on the jobsite, the coordination of the expectations must become a part of the process. As a part of the process the system expectations and actions become predictable and reliable, so the system performance and output also become more predictable and manageable. The types of issues that should be considered in the design of the process at this level are:

- Location of delivery by vendors or third party carriers
- Manpower required to unload trucks and who will provide the manpower
- Types and quantity of equipment needed to move material on the jobsite and who will provide or arrange for the equipment
- Procedures to deal with delays and changes after the material has been delivered
- Process exceptions and variations that will be needed to accommodate errors and other emergency material needs

The design on the process accommodating these factors has to be keenly focused on the total procurement cost within the supply chain. In this area there are many historical paradigms that need to be challenged. A thorough understanding of costs, costs that have historically not been tracked will be necessary to truly optimize the division of work and compensation in this area. The expertise exists and the experts are known and recognized; however, the traditional accounting measures for cost do not allow for the readily available Activity Based Costing (ABC) model for vendor services. At this point these factors must be resolved through negotiation, trust, and refinement.

Step 5: Returns

Figure 5.9 outlined the rate of material ordering and delivery in contrast to consumption and installation showing the returns in the form of ordered and delivered material in excess of 100 percent. The procurement process must recognize several characteristics of returns, the most important of which is that everyone pays dearly for every returned item. As a rule of thumb, the contractor should only order extra material if he is willing to discard the unused items. This is not to suggest that returns should never be processed and will not occur; they will. The goal has to be to improve the entire procurement process such that inventory management and planning eliminates the need for costly buffers, and consequently the need to divest the jobsite of these buffers on completion. The simple fact that inventory levels are being reduced will inherently reduce the quantity of returned goods. Figure 5.4 shows the cost to the distributor for handling and processing returns. This cost is real and is included in the selling price of every item that the contractor purchases. Reducing the returns will directly lower the distributor's operating cost, and will ultimately translate into lower material pricing for the contractor.

Step 6: Lessons Learned and Postmortem

There are several levels of learning involved with the procurement process. Truly creating a continuous improvement methodology in a process that involves several stakeholders will require dedication more so than careful planning. It is often less important how the lessons are captured and communicated than the simple fact that they are captured and communicated in a timely fashion. Of course, converting the information into wisdom and system improvement requirements will be driven by how the shared information is used within each organization and within the system as a whole.

At the simplest level, learning has created the current situation. This is the result of learning without the presence of a standardized or formal procurement process. Foremen have observed the issues and the failures of the current procurement model and have modified their own behavior to accommodate their experience. For example, foremen who are consistently plagued with back orders and incorrect material will tend to order more than needed, or even for the entire project at the beginning and feel more comfortable in the fact they have enough to work his/her Plan A, Plan B, and even Plan C if necessary. In this case foremen will typically want everything sent immediately because if the material is not onsite, then they are concerned that it will not appear. This is a system that involves excess material inventory that is ordered with little or no lead time (next day orders for large quantities of materials).

By communicating and sharing his/her observations of the system's failures, the foreman can allow solutions to be developed and implemented at the source. Rather than accepting the inefficiencies and paying for them, the system can improve its performance and eliminate the errors as well as the cost associated with these errors. Only with proper timely feedback can the system be improved to reduce the cost of poor quality. Achieving this level of correction requires sharing and learning beyond the jobsite and the contractor's office; the learning will have to extend to include the entire supply chain and down to the general contractor and to the customer.

An effective issue identification, communication, and resolution component is a vital ingredient in any complex business process. It is foolish to believe that anyone can develop a perfect process the first time. It is equally foolish to assume that what worked last time will continue to work forever. The procurement process must have the ability to be modified with the benefit of learning from lessons captured by all stakeholders and communicated throughout the supply chain.

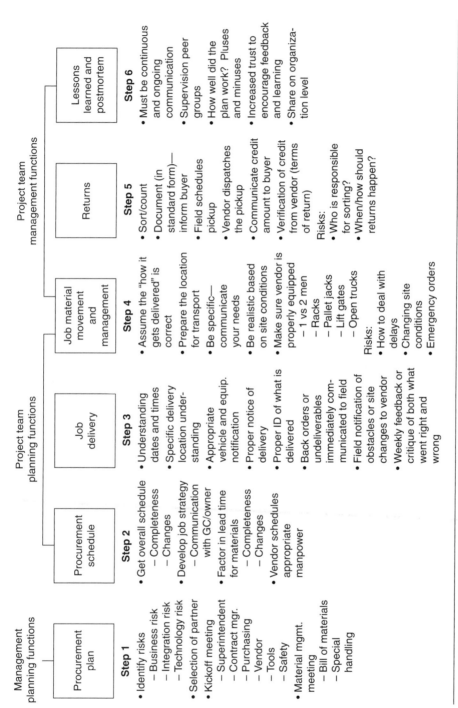

Management planning functions

Procurement plan

Step 1

- Identify risks
  - Business risk
  - Integration risk
  - Technology risk
- Selection of partner
- Kickoff meeting
  - Superintendent
  - Contract mgr.
  - Purchasing
  - Vendor
  - Tools
  - Safety
- Material mgmt. meeting
  - Bill of materials
  - Special handling

Project team planning functions

Procurement schedule

Step 2

- Get overall schedule
  - Completeness
  - Changes
- Develop job strategy
  - Communication with GC/owner
- Factor in lead time for materials
  - Completeness
  - Changes
- Vendor schedules appropriate manpower

Job delivery

Step 3

- Understanding dates and times
- Specific delivery location understanding
- Appropriate vehicle and equip. notification
- Proper notice of delivery
- Proper ID of what is delivered
- Back orders or undeliverables immediately communicated to field
- Field notification of obstacles or site changes to vendor
- Weekly feedback or critique of both what went right and wrong

Job material movement and management

Step 4

- Assume the "how it gets delivered" is correct
- Prepare the location for transport
- Be specific—communicate your needs
- Be realistic based on site conditions
- Make sure vendor is properly equipped
  - 1 vs 2 men
  - Racks
  - Pallet jacks
  - Lift gates
  - Open trucks

Risks:
- How to deal with delays
- Changing site conditions
- Emergency orders

Project team management functions

Returns

Step 5

- Sort/count
- Document (in standard form)—inform buyer
- Field schedules pickup
- Vendor dispatches the pickup
- Communicate credit amount to buyer
- Verification of credit from vendor (terms of return)

Risks:
- Who is responsible for sorting?
- When/how should returns happen?

Lessons learned and postmortem

Step 6

- Must be continuous and ongoing communication
- Supervision peer groups
- How well did the plan work? Pluses and minuses
- Increased trust to encourage feedback and learning
- Share on organization level

Figure 5.11 Complete detailed process of procurement

Finally, at the highest level of learning, there is opportunity for contractors to make gains industry wide. Through the formation and expansion of peer groups and a productivity center mentality it becomes possible for contractors to benefit universally from lessons learned by one another. Because of the large labor expense associated with material handling this is the single greatest focus area for contractors to improve productivity and profitability.

In summary, the complete six-step procurement process that has been outlined in this section is the key to implementing the results of this research. The contractor who desires to minimize their jobsite inventory can only accomplish effective operation at, or near, the ideal level with a collaborative business model that leverages the capabilities and the resources of all stakeholders in the project. The complete and detailed process is outlined in Figure 5.11. This process can be used to act as a road map, or a guide to developing a specific application of these concepts.

CONCLUSIONS

The objective of this study was achieved and the ideal level of jobsite inventory has been identified to be three days for both the current procurement model and potentially for an improved model. Reductions of jobsite inventory from the current industry norm of more than two weeks, to an immediate target of three days and ultimately to an ideal level of approximately one day will require significant changes in the practices of the construction industry, not just a decree from management. The data have shown that the current understanding and stated material procurement policies of the industry are being followed because the system is not currently capable of meeting this demand. Achieving the ideal inventory level at the jobsite without disrupting and further eroding profitability will require a total systems approach to material procurement and management for construction.

To advance the competitiveness of this industry further, results of this study need to be implemented through workshops and seminars or direct company-by-company application. The results of this research suggest that an industry-wide forum needs to be created. This forum can help develop industry, business, and project level procurement models and educational modules for contractors as well as for their vendors and manufacturers. These new procurement models will need to be developed and implemented throughout the industry to have a significant impact on the total productivity of construction.

6

TOOL AND MATERIAL CONTROL SYSTEMS

Dr. James E. Rowings, *Kiewit Corporation*
Dr. Mark O. Federle, *Marquette University*

INTRODUCTION

Keeping track of tools and materials is a major problem for contractors. The results of haphazard control systems are (1) higher cost to replace lost equipment or tools, or buying unneeded materials, which might only be "buried" in the warehouse and (2) higher personnel costs due to less efficient use of staff to remedy problems that could be solved in another way.

The key issue for contractors implementing tool and material control systems is the cost and burden of the systems themselves versus the cost benefits to the company of improved control (minimizing tool cost and the unnecessary purchase of tools and materials).

There are also intangible benefits of improved tool and material control, which are much harder to estimate with confidence. For example, it is difficult to measure the impact of improved efficiency by craftsmen in the field because they have the tools and materials needed to support their work.

The management techniques explained in this chapter are intended to assist contractors in improving their operations. The most important item to keep in mind is that either a tool or a material control system should have demonstrated benefits that exceed its costs. These benefits may be in increased productivity of the craftsmen in the field.

With the exception of start-ups, all companies have tool and material control systems, even if they are "do-nothing" systems. The decision to update or implement new systems is typically prompted by dissatisfaction with the adequacy of the current system. Many different methods are used to control tools or materials within the construction industry. All fall into two major categories: manual and computerized. Factors influencing the type of system chosen include company size, number and dispersion of jobs, and number of tools and materials to be controlled.

Manual control systems generally fit smaller contractors best because the number of tools and material transactions is limited. More sophisticated computerized systems make sense for larger firms with much larger quantities of tools and materials to control.

Ultimately, people make a system successful. A poor system can be made to function by having dedicated, trusted employees, who are committed to efficiency and control of company tools and materials. The most exacting control system, on the other hand, will fail without the commitment and dedication of the people responsible for implementing it. Users must believe that the system will contribute to the bottom line of the company's operations—and thus ultimately to their own prosperity and security.

Successfully implementing tool and material control systems requires commitment and buy-in at all levels of the company. Involving people at all levels in the design of the system is the best way to ensure acceptance.

TOOL CONTROL SYSTEMS

Opportunities for Improvement

Establishing a more effective tool control system depends on many factors. Basic techniques for improving tool supply to jobsites include:

1. Tool thesaurus
2. Tool control manager
3. Prepackaged tool sets
4. Start-up lists for typical jobs or activities
5. Voice mail for tool requests
6. Jobsite accountability system for tools
7. Bar-coding systems for tools
8. Tracking everything for the psychological effect
9. Strengthening communications
10. Standardizing tool brands

Tool Thesaurus

A tool thesaurus is a reference book of company-specific tools, possibly with pictures, that allows office support staff to understand what the foreman or craftsworkers are requesting when they phone in a tool request. Such a book will prevent the field-worker from calling back if the tool control person is busy or away from the telephone.

Tool Control Manager

There must be a single knowledgeable person in charge of the tool function, to ensure control and to assist foremen in identifying methods of accomplishing the work tasks.

Prepackaged Tool Sets

Some contractors use prepackaged tool sets for repetitive work, similar to standard sets of tools in a service vehicle.

Basic Start-up Lists

Start-up lists for typical jobs and activities assist proper planning and reduce or eliminate time lost due to lack of tools at the jobsite.

Voice Mail

Using voice mail for tool requests avoids lost time and confusing requests due to miswritten messages.

Jobsite Accountability for Tools

Monthly tool reports classified by jobs or a review of tool costs for each job will increase foremen's accountability for ensuring the tools are not lost or misplaced.

Bar-Coding Systems for Tools

This may provide the level of control needed by certain contractors.

Tracking Everything for Psychological Effect

Many contractors divide tools into capital and expendable categories, and concentrate primarily on tracking the more expensive capital tools. However, some contractors prefer tracking every tool they own to develop a psychological climate of honesty and responsibility.

Strengthening Communications

Improving communication between those who currently have tools, those who need tools, and those in charge of managing tool supply may help reduce the incidence of tool hoarding.

Standardized Tool Brands

For many types of standard tools (e.g., drills, saws, pipe benders) standardizing with a single manufacturer eases maintenance and tracking.

General Questions

The two general categories of tool control systems are manual and computerized. Manual systems may be more appropriate for small contractors. Computerized systems are usually more appropriate for large contractors. Any contracting firm just beginning to implement a tool control system, or improving a current control system, should begin by answering the following general questions. They are intended to help with decision-making, regardless of whether you choose a manual system or a computerized system. Discuss these questions with key managers, foremen, and craftsmen to ensure adequate input is received from all parties who will be involved in making the control system a success. This will help ensure buy-in from all parties and increase the likelihood of success for your tool control system.

What Tools Will You Track?

Is there a minimum value (cost) associated with a tool to determine whether to track it or not? Will you track expendable and capital tools in the same manner? If not, how will the system you choose provide you with the necessary flexibility?

Who Will Be in Charge of the Tool Control System?

Who will be allowed to transfer or send tools out to a job? Will you allow transfers directly between jobs or will they first have to come back to tool storage?

Who Will Determine Whether a Tool Is Working Properly?

Some contractors rely on the workers on the job to tell the tool control manager when a tool is in need of repair (typically by using repair tags). Others test or operate each tool before sending it back into the field.

How Will You Establish an Indexing System?

Chances are you can use the index of your largest tool distributor for a start. You will then have to establish an indexing system for separating tools by categories, types, and individual tools.

Who Will Conduct the Initial Inventory?

A single individual (the tool control manager) is the best person for this task. Once you begin the inventory and tagging of tools, you must also begin the process of formalizing the tool transfer procedure. Thus, make sure the procedure is in place prior to beginning the inventory. Note that much of the inventory will necessarily be conducted on the various jobsites the contractor is currently working on. Once a job has been inventoried, it must be treated as if the control system is already operational (completing transfer forms, and so on). Otherwise, maintaining an accurate inventory will not be possible.

How Can You Make Your Tools Unique?

If you have not done so previously, an ideal time to mark your tools to make them stand out on the jobsite is during inventory. This prevents loss due to other trades working in close proximity to craftsworkers. Tool marking methods include brass tags, bar codes, and painting tools in bright colors.

Tool Control Manager

The key person in any tool tracking system is the tool control manager. This must be a knowledgeable person who can help foremen determine which is the best tool for a particular job, maintain and repair tools as needed, authorize purchase of new tools, and manage movement of tools among jobs. The tool control manager also ensures that all the pieces needed to operate a certain tool are shipped to the jobsite together. In many companies, the person designated as tool manager may have other duties as well.

Classification of Tools

Before establishing a tool control system, a contractor will need to decide whether to include only capital or capital and expendable tools within the control system. Capital tools are those items that are typically depreciated (for accounting purposes) over several tax years and may be used on many jobs. Expendable tools are those purchased for a particular job or as overhead, and which are expensed (for accounting purposes) in the current tax year. While you may not track maintenance and repair for an expendable tool, you may want to know its location on which jobs or in which service tracks.

These are not firm classifications within the construction industry. Some contractors may treat a ⅜″ drill as a capital tool, while others may consider only a tool with a replacement value of $2500 as a capital item.

Tool Tracking Systems

Once the evaluation of the system choices (manual, computerized, or both) and the inventory of the tools to be included have been completed, a tracking system must be developed. Figures 6.1 and 6.2 present a flowchart for the decisions and activities surrounding the tool tracking system. This tracking system consists of a unique number for each tool and a method for tracking usage and repair costs. Many contractors use brass tags or etch these numbers directly on the tool. An example of one code must be as follows: ⅜″ Drill-08-05-01. This would be used to track the first (01) 3/8″ drill bought during May (05) for 2008 (08).

Proper documentation is critical for an accurate inventory and tracking of tools. The tool control manager should input these data to ensure accuracy. While it may be tempting to use an assistant to input data, the cost penalties associated with inaccurate data or entry verification are much greater than the minor savings realized by having an assistant enter the information. Once the tracking system is set up, an assistant may be able to maintain the system inputing monitoring information under the general supervision of the tool control manager.

Tool purchases typically occur in response to a foreman's request when similar tools are not owned or are unavailable. Figures 6.3 and 6.4 present a flowchart of tool control activities. Once the foreman's request is received, the tool control system is checked to determine if that tool is available.

For some contractors, the process of determining whether a tool is available is fairly laborious. A warehouse person calls around to each project to determine if the other foremen really need that particular tool. This time-consuming process can be reduced significantly if the warehouse staff knows which jobs have which tools. If the tool is not available, a new tool must be purchased, coded, and entered into the tool inventory and control system. The tool is then ready for transfer to the field.

Once a crew is done using a tool, it should be returned to the warehouse. Direct transfers between jobs should not be tolerated, since tracking tools under direct transfer is likely impossible or extremely difficult. Hoarding popular tools also should not be tolerated.

When tools are returned to the warehouse, the condition and serviceability must be determined. There are two common methods for this determination: (1) the foreman separates functional or safe tools from those requiring repair, which then are "red tagged" with the problem description. This method depends on crew members to report tool repair needs to their foreman, who then must remember to complete the "red tag" and (2) warehouse staff performs a functionality test and safety inspection of each tool when it is returned. Although this approach is more laborious for the warehouse staff, it ensures that each tool sent out to the next job is in working order.

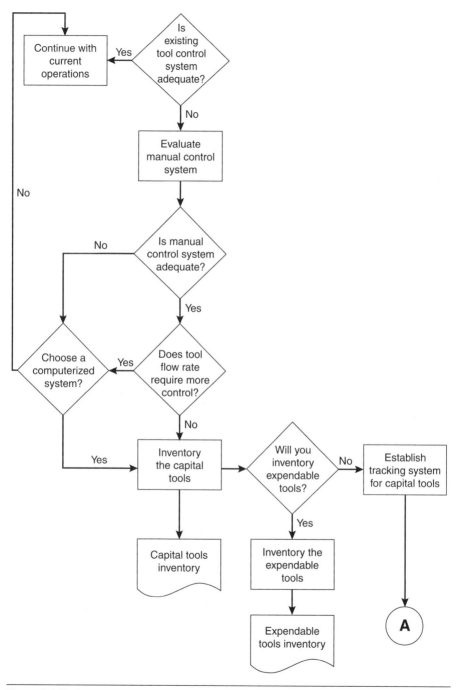

Figure 6.1 Tool control selection and implementation flowchart (page 1)

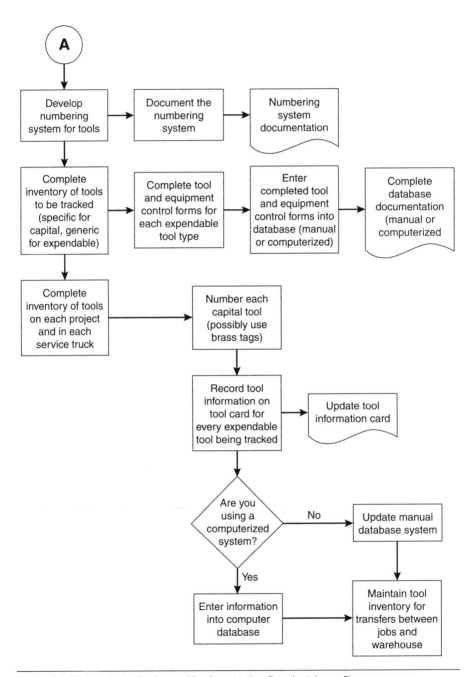

Figure 6.2 Tool control selection and implementation flowchart (page 2)

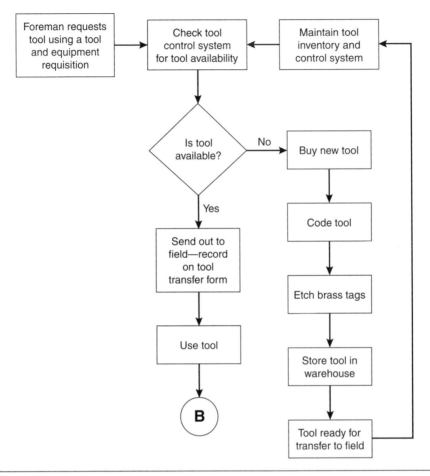

Figure 6.3 Tool control flowchart (page 1)

While a tool is being repaired, the tool inventory must be updated to reflect its status and lack of availability for assignment. When the tool is ready to return to the field, the tool control database should be updated to reflect its availability. The maintenance record on the tool should also be updated to reflect the repair work performed.

Some contractors repair many tools in-house. When a tool cannot be repaired cost-effectively, then it is used for parts. At this point, the tool should be removed from the tool inventory and added to the parts inventory (with a listing of those parts that are usable).

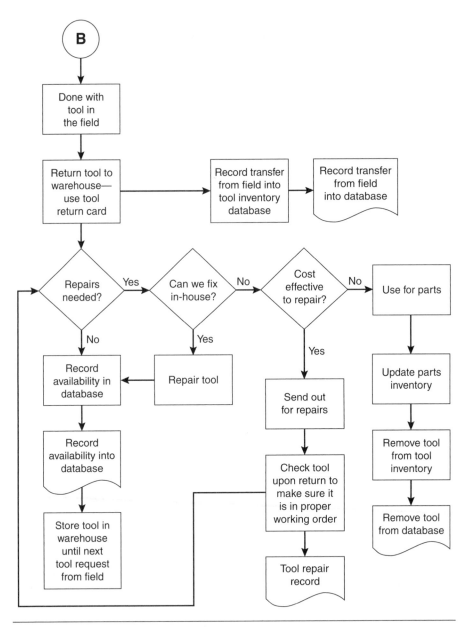

Figure 6.4 Tool control flowchart (page 2)

Manual Tool Control System

Small contractors may not have the financial capacity to afford a large capital expenditure to track tools. Manual control systems are most appropriate for small contractors due to their low cost and simplicity. The system described in this section is based on an actual system used by a small electrical contractor in Phoenix, Arizona.

The initial cost of this sample system is quite low, typically a few hundred dollars. The materials needed are a piece of plywood, finish nails, color-coded washers, brass tags, and an engraving tool. The operational costs, however, are significant and require a trusted, dedicated tool manager familiar with the contractor's work and the operations of the tools used.

This system consists of two tracking boards, one for expendable tools and another for capital tools. Separate columns for jobs currently under construction (or service trucks) run across the top of each board. Each job name is typed or handwritten on a card and inserted into a cardholder to allow for changing projects. A list of tools to be tracked runs from top to bottom on both sides of each board. Colored washers are used for each individual tool. If Job XYZ had four long-handled, round point shovels, four washers would be placed on the single nail that identifies the round point shovel and Job XYZ. Using colored washers to represent each expendable tool provides the contractor with a method for tracking and knowing where all tools are. This method allows them to control the movement of tools among different jobs, and thus reduce the number of new tools purchased because none were back in the warehouse.

If you decide to implement a similar system, you must first determine the size for the tracking system needed by your company. To accomplish this you must calculate the typical number of ongoing jobs and service tracks in operation at any given time, allowing some room for company growth. You should probably allow enough columns across the top of the board to list three to four times as many potential jobs (and service trucks) as your current total jobs and service tracks.

To determine the number of rows needed down the sides of each board, list all the tools that you and your craftsmen currently use and for which you want foremen to be accountable. To allow for future growth, make twice as many rows as the number of tools you will track at this time.

Remember that what gets measured gets done. In tool control, this means that tools, which are tracked, are more likely not to be lost, stolen, or misplaced. It is important that the only person who can put washers on and take washers off the tool tracking board is the tool control manager. If authority to handle washers is given to multiple people, there is no practical way of tracking tools.

Implementation of a tool control system requires commitment and discipline by everyone involved. The success of the approach described in this section by the contractor in Phoenix is based on a strong commitment to tool control by the company's top management. To maintain the integrity of a tool control system over time requires discipline to make the system function as desired at all times with no exceptions. Control is only possible with this discipline and attention to cost-effective tool repair and replacement decisions.

This contractor uses an engraved brass tag system to track capital tools. Each tool is uniquely identified by the information on the tag. Each tag uses a date code to identify how old the particular tool is, and thus enables quicker repair and replacement decisions. These tags are placed in the nails as explained earlier, according to the tool type and the particular job or service truck. A duplicate brass tag is made and permanently attached to the tool or the tool case. Engraving is used to mark capital tools when there is no way to permanently attach a brass tag. In addition, the contractor also uses a card system to maintain information regarding each capital tool, including date purchased, serial number, and maintenance history.

A manual tool control system should be kept as simple and as practical as possible. The goal is to know where your tools are and to minimize the unnecessary purchase of new tools. The typical small contractor does not need to invest in computer technology for tool tracking. As a company grows, however, the increasing numbers of tools and jobs may outgrow the limitations of a manual tracking system. At this point it makes sense to start looking at computerized tool control systems.

Computerized Tool Control System

A computerized tool control system can be established from scratch or an existing manual system can be converted to a computerized one. In either case, there are six steps to the process:

1. Choose the correct software for your business
2. Decide which tools to track
3. Locate those tools
4. Code those tools
5. Enter the data
6. Control on an ongoing basis

When considering the move to a computerized system, make sure that you accurately consider the costs of supporting and updating the control system. The accuracy of the tool control system relies on consistent daily updating of the tool transfers between jobs. Also, future markets may require flexibility in the system that must be recognized during the purchase decision process.

If your tool control manager is not a computer literate person, he or she will need support from your company for training. The tool control manager must be extremely knowledgeable about the tools and their repair; this person should help foremen decide which tools they need and whether a tool that is available could meet their current need. The software system will simply assist this person in locating and controlling your tools and thus hopefully reduce tool acquisition costs.

A computerized tool control system can be implemented using three types of software: (1) spreadsheets; (2) databases; and (3) commercial tool control programs. Each approach has advantages and disadvantages.

The tool board and brass tag system described in the previous section lends itself to a spreadsheet application. The main advantage of using a spreadsheet program is that your company might already use spreadsheets for other tasks. The greatest drawback in using a spreadsheet is that developing links between the various workbooks and maintaining a single master copy may prove to be difficult.

A more sophisticated approach can be developed using an off-the-shelf database system. A strong advantage of the in-house-designed database system is the flexibility that the system will provide. Because you design it, you can track your tools however you want. While many commercial software packages will claim to provide the ultimate flexibility, only a system designed completely in-house will truly provide you with that luxury. The main disadvantage of developing your own database system is that your in-house costs may be higher than the costs of a commercially available package.

The easiest and probably fastest method of computerizing your tool control system is to purchase a commercial software package. There are several commercial packages available in the market. The software you buy must fit your company's individual needs. Be thorough in your evaluation of each software package. Some factors to consider include management issues and software features as explained in the following sections.

Management Issues

1. Software technical support—Is there a number you can call? If so, call it—see how long they put you on hold.
2. Upgrades—Are they constantly revising the software? This may indicate enhancements to the software, which are normal. Or it could indicate the need for fixes to "bugs" within the software, which is cause for concern.
3. References—What names do they provide as references? You probably do not want to be their largest client.
4. Training—Do they offer training? This may be a sign that their software is difficult to learn. If you experience turnover within your company, the training of new people on the software might be of concern.
5. Hardware compatibility—Is your existing computer system adequate to run the software and fast enough to provide quick and easy access to the tools you are trying to track?

Software Features

1. Compatibility with your company—If you have a manual system, will the software allow you to build from the manual system? Or will you have to change the way you track tools? Are you willing to change your company's system because of a software package that is supposed to help you?
2. Capacity—Can the system track thousands of tools, using a variety of types and descriptions? Can the system track both jobs and service vehicles? How expandable is the system?
3. Reports—Can you generate the reports that you want? Report examples include
   - Tools assigned to an individual (such as personal safety equipment).
   - Tools on each job (comparing personnel to number of tools lets the tool control manager determine whether a foreman is "hoarding" tools).
   - Tools in each service truck.
   - Complete inventory list of all tools.
   - Can the cost of these tools be included in the same report, or will you have to create several reports to have the information you need to control?
4. Maintenance—Can you track the maintenance and repair history for an individual tool and a tool type? This may be helpful in making repair and replacement decisions.
5. Transfers—Can you generate a history of where the tools have been (transfer log)? How time consuming is it to transfer a tool (or an entire crib) from one job to another?
6. Rental rates—Can rental rates be assigned to tools, and rental rate reports be generated to assist service persons and estimators in costing out a job? Can this be completed for different daily, weekly, and monthly rates? Can a tool type be updated for its rental rate or does each individual tool need updating (much more time consuming)?
7. Lease tools—Can the software package track long-term leased tools, identifying them differently from company-owned tools?

MATERIAL CONTROL SYSTEMS

Opportunities for Improvement

Material control has a major impact on contractors' bottom lines. Establishing a more effective material control system depends on many factors. Basic techniques for improving material supply to jobsites include:

1. Material handling improvements
2. Warehousing practices
3. Credit practices
4. Online supplier inventory checking
5. Waste control
6. Work assemblies
7. Staging and pre-planning
8. Bar codes/electronic data transfer

Material Handling Improvements

Some contractors negotiate a return policy to reduce the cost impact of hidden freight damage. They may refuse to accept back orders from their partnered suppliers.

Warehousing Practices

Reorganizing your material storage warehouse may lead to improvements in locating materials and allow these materials to be used on future jobs (as opposed to new materials being purchased).

Credit Practices

Negotiating a change in credit practices may improve cash flow.

Online Supplier Inventory Checking

You may be able to check your supplier's inventory at their websites directly from the field through personal computers or smart phones.

Waste Control

Changing the packaging of some materials may reduce the disposal costs of packing materials and the time spent to unpack.

Work Assemblies

Organizing work into assemblies may allow palletizing certain groups of materials and organizing them according to the order in which the materials will be put in place.

Staging and Pre-planning

Similar to assemblies, staging and pre-planning may reduce the time spent looking for the needed materials.

Bar Codes/Electronic Data Transfer

Bar codes might also be used to track materials and report upcoming material needs to the home office.

General Questions

Figure 6.5 illustrates a traditional approach to materials management. This figure shows the large number of factors that impact the overall materials management system. For most contractors, all these items will be the responsibility of one or two persons.

Even with a small number of personnel involved, it is nonetheless critical that a materials management plan be developed. Developing a plan requires answering the following types of questions. There may be other questions as well, specific to your own company's size, structure, and management style:

1. Which companies will be our primary suppliers?
2. Will we warehouse materials?
3. How will access to the warehouse be controlled (especially critical for contractors with service fleets)?
4. Who can issue purchase orders (POs) (are POs always required)?
5. Will we track the quality of materials supplied?
6. Will we track the accuracy of material quantities supplied?

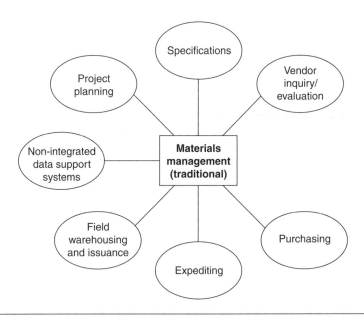

Figure 6.5 Traditional materials management functions

Material Control Flowchart

The material control flowchart is intended to be used both by those contractors who warehouse materials, and by those who order materials directly from a supplier and return the surplus to the supplier. The following is a general description of the multistep process outlined in Figures 6.6 and 6.7:

1. Bill of materials—develop a bill of materials. This will likely build off of the estimate, but must be in much greater detail than a typical estimate would be.
2. Inventory—determine which of the needed materials is already available. The stock of warehouse materials for the job is placed in the job trailer or shipped to the job. The inventory is reduced accordingly.
3. Additional materials—order additional needed materials from suppliers and have delivered to the jobsite. As the project nears completion, a determination of what additional materials are needed is required. These can either be delivered from the warehouse or from a supplier.
4. Remaining materials—determine those materials to dispose of, those to store, and those to return to a supplier. After the project is completed, the inventory must be updated when materials are returned to the warehouse.

Materials Management

Warehousing and Inventory Management

Like a tool control system, a material control system can be either manual or computerized. The initial steps are the same in either case. Developing a material control system begins with identifying the materials needs for a project. Typically, a materials order list would be developed by using the estimate as a starting point, and then adding significantly more detail.

A manual material control system depends on a well-organized warehouse. Especially for contractors running service organizations, some standard items may always be stocked, while many of the leftover items from previous jobs will remain until they are used in a future job. One downside to stockpiling leftovers is that money is tied up in inventory. Only at inventory time will you know the extent of this investment and perhaps the availability of certain items.

A computerized material control system allows the contractors to more easily track large amounts of information. For example, it allows estimators to search through a database for current inventory, rather than wandering through

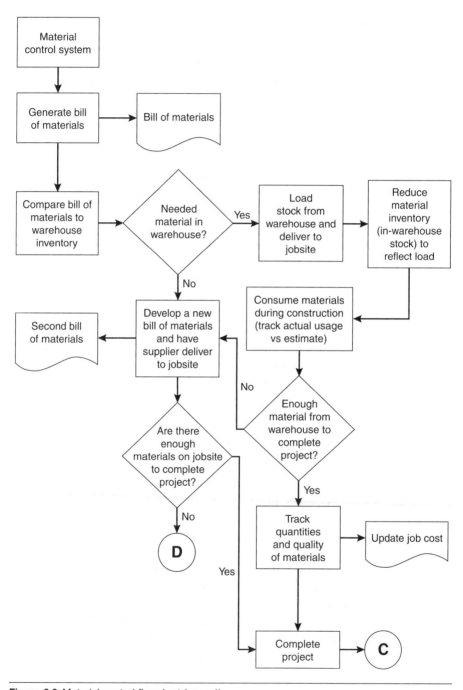

Figure 6.6 Material control flowchart (page 1)

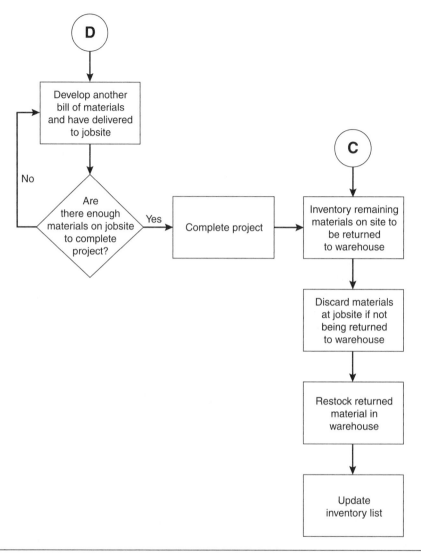

Figure 6.7 Are there enough (page 2)

warehouses looking for a particular item. Timely information regarding current inventory value, the number of turns, by item and overall, the overstock, and the back-order information are all readily available.

Developing a material control system begins with the following pre-planning and scheduling steps:

1. Analyzing the requirements
2. Pinpointing the critical items

3. Relating particular materials to specific sections and sequences of the installation
4. Completing the estimate list (filling in the holes, defining materials needs more accurately)

There is a great deal of information regarding materials management available in the construction industry. The Construction Industry Institute has established several basic premises of material management:

1. Productivity—materials management is a clearly defined task that, when properly planned and executed, provides project management with an invaluable tool to optimize schedules and improve labor productivity.
2. Savings—substantial savings accrue to both owners and contractors. Some studies estimate these savings at six percent of project craft labor cost.
3. Support—a material management system requires top management support to be successful. A major hurdle that must be overcome is the resistance to accept the up-front visible costs of a well-planned materials management effort. However, this cost will always be less than the hidden cost of unplanned material handling and control.
4. Training—training is an essential element in successfully implementing a material control system. Training must include not only the materials management personnel, but users at all levels of the project team.

Purchasing

Virtually all jobs require material purchases from a supplier or multiple suppliers. The two major methods of purchasing materials are by a blanket PO-release shipment and by individual POs.

In a blanket PO-release shipment, the quoted material prices are included in an early blanket PO, and material is requested (with separate release orders, referencing the original blanket PO) as the job progresses. This approach allows use of a single PO for the job, but also allows tracking and checking of materials on each release order. This method is typically used if the jobsite does not have large spaces for material storage.

In individual POs, a separate PO is issued each time material is required by the jobsite. Material procurement will typically be completed by the jobsite foreman or the home office project manager. This approach builds off the project schedule; material purchases are paced to allow just-in-time delivery.

Material Returned to the Warehouse

Just as with the tool control system, if a material control system is put in place, there are two primary options: a manual or computerized system. However, since it is likely that accounting is already computerized, and much of this information is needed to issue invoices and check incoming invoices, data entry duplication may be avoided by computerizing the material control system.

All materials needed for the job, as identified from the material schedule, would be shipped to the jobsite from the contractor's warehouse. To adequately track the flow of materials within a contractor's warehouse, material requisition and return material forms must be used. There are several methods of controlling leftover materials from a completed jobsite:

1. Discard—some contractors simply throw away leftover materials at the end of a job, to avoid the trouble and cost of managing it. These contractors assume that the cost of purchasing new materials for the next job is less than the cost associated with returning the materials to a central warehouse, maintaining an inventory control system, and delivering the needed material to the next project.

2. Return to supplier—some contractors negotiate a restocking fee with their suppliers and return material acceptable to them (typically unopened cartons and undamaged fixtures), scrapping the rest.

3. Return to warehouse—Figure 6.8 presents a flowchart of returned materials. This would allow materials already owned by the contractor, which might be put into a bid at the original cost or reduced price, to be spoken for by a particular estimator. An internal "restocking" fee would be charged for materials returned to their warehouse to help them job-cost appropriately those expenses associated with restocking, carrying, and maintaining the inventory.

4. No credit for returns—an interesting approach to reduce materials returned to the warehouse is not to credit the original job for returning this leftover material. This allows a contractor to job-cost the material once plus the number of times it is returned to the warehouse, helping reduce the warehouse expense. This "no credit" method also keeps foremen and superintendents aware of the costs being charged to their jobs and may reduce costs by preventing the ordering of too many items that later have to be restocked.

Job Billings

Just as important as tracking the materials is tracking the paperwork associated with the materials, whether ordered from an outside supplier or received from

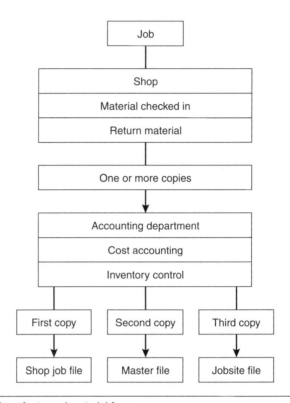

Figure 6.8 Routing of returned material forms

the contractor's own warehouse. Making sure the material requisition reaches its proper destination in a timely way is important to ensure proper job progress, and ultimately proper job-cost billing.

The last area of concern for a material control system is how to properly credit the job for material returned to the warehouse or supplier. The flowchart in Figure 6.8 shows that the material return form is used to properly credit the job from which the material was returned and also allows maintenance of an inventory (or material control system) by using the copy returned to the shop job file.

Computerized Material Control System

A computerized material control system can be extremely valuable for a contractor. However, there are several key aspects that contractors should consider when developing a computerized system, among them:

1. Can the program track information for several thousand items? Are additional items easy to add later?
2. Is the level of information detail adequate for your needs? This may include cost, multiple levels of pricing, minimum quantities to be stocked, maximum quantity, reorder lot size, or lead time.
3. Can inventory be cost the way you want (e.g., last-in-first-out (LIFO), first-in-first-out (FIFO), weighted average, set cost, most recent cost)?
4. Does the program allow you to track inventory to multiple types of sites, including trucks, projects, warehouses, and yards?
5. Can you hold inventory for upcoming jobs?
6. Is the inventory system tied to job cost so that it is updated automatically to reflect the requisite material expense?
7. For report generation, can reports be displayed on a screen, saved to disk, and printed as hard copy? Can the inventory be sorted by site, type, description, or by preferred (primary or secondary) vendor?
8. Can you quickly determine what your current inventory levels are? Can you determine what your current and year-to-date activity is, as well as cost, turnover, and gross margin?
9. If you also have service billing, are those two systems integrated?

While contractors need a control system, it should be one of their own choosing. Do not be influenced by a software salesperson to choose a system that you neither want nor will you use. A tremendous amount of software sold to construction companies across the United States barely comes out of its shrink-wrap.

CONCLUSIONS

Regardless of the size of contracting firms, there is an obvious need for them to control their materials and tools. Contractors are throwing away profits that could end up in their own coffers by not improving their methods of controlling tools and materials. This chapter presents a number of flowcharts and methods for controlling tools and materials. Review each of these before choosing a method of control for your company. Only by custom designing your control system, will you be satisfied with the result. As we have learned from successful management processes, what gets measured gets done. If you do not do anything to control your tool loss or material overages, you will always be frustrated with additional and unexpected costs.

The management techniques explained in this chapter are intended to assist contractors in improving their operations. The most important item to keep in mind is that either a tool or material control system should have demonstrated benefits that exceed its costs. These benefits may be in reduced losses or in increased productivity of the workers in the field.

7

RECOMMENDED SAFETY PRACTICES

Dr. Clark B. Pace, *University of Washington*
Dr. Eddy M. Rojas, *University of Washington*

INTRODUCTION

During periods of decreasing economic prosperity, companies of all sizes tend to implement strategies to bolster their financial position. There are many strategies to improve profitability. Some companies may choose to adjust their marketing to attract different clients while others may focus on cutting costs. Contractors may cut staff or programs that are perceived as expensive or unproductive.

The worst possible mistake a contractor could make would be to cut back on a safety program. Not only do safety programs protect the health and welfare of employees, they significantly impact profitability. The safety rating of the company significantly affects profitability, regardless of the size of the company or the nature of work it performs.

The reason that any contractor is in business is to turn a profit. As profit margins decline due to a slowdown in the number of new projects under construction and an increase in the number of contractors looking for new work, companies are looking for creative ways to attain a competitive edge. However, a company that looks only to cut existing costs may miss a potential savings that may be realized by learning to prevent future costs. A prime example of this mindset has to do with safety. Besides the obvious and quantifiable costs associated with an incident, there are some costs associated with a poor safety record that might be overlooked.

SAFETY AND THE BOTTOM LINE

Safety affects the "bottom line" in one significant way, as well as in a number of peripheral ways. The major impact of safety is through the experience modification rate (EMR). A company's EMR is a means of measuring its overall safety performance versus an industry standard. The EMR is a multiplier that is used to adjust the amount of money a company pays to the governing body each man-hour toward workers' compensation insurance. Company A with an EMR of 0.70 essentially receives a 30 percent credit toward its worker compensation costs. Company B with an EMR of 1.30 has to pay a 30 percent surcharge for its premiums. The difference is certainly significant; Company B pays nearly twice as much per hour per employee than does Company A. All other things being equal, Company A will always be able to offer services for a lower cost than Company B because it effectively pays less for its labor force. EMR as a safety evaluation tool can be problematic, in that there are many instances wherein a company will pay an employee to not file a workers' compensation claim to prevent its EMR from rising. Also, more and more companies are providing their own insurance, as opposed to state-run workers' compensation insurance programs, so EMR is no longer an applicable measure. Further, EMR can be biased against small companies, in that a single claim for a small company can result in a dramatic rise in its EMR, whereas a large company may have several claims but still maintain a lower EMR.

Other, less tangible effects of safety are much harder to quantify, but may still play a role in the overall profitability of a company. Owners do not want the liability or negative publicity of having a project with negative safety incidents. One need only read about the public backlash that occurred after accidents on such high-profile projects as the Alameda Corridor Project or the construction of the new baseball stadium for the Milwaukee Brewers to understand this. Owners can and often do use safety as a pre-qualification criterion when choosing a contractor. A poor safety record may prohibit a company from even bidding or negotiating a project. Safety impacts a company's ability to compete for work.

Safety also has an effect on the morale and ultimately the productivity of field employees. For example, a craftsman who works for a company with a poor safety record, or that places a relatively low value on the safety of its employees, may choose to look elsewhere for another job. In the long run, this means that an employer will constantly be forced to train new employees, and the employees on any given job will always be on the steep part of the learning curve. Productivity and quality may suffer as a result. On the other hand, if that same craftsman feels that he works in as safe an environment as possible, and that his employer cares for his well-being at work, then that individual and his coworkers may be able to increase their productivity through increased job satisfaction, proper planning, and understanding of project requirements. Safety improves worker productivity.

CAUSATION THEORIES: WHY DO ACCIDENTS HAPPEN?

When discussing safety issues, employers often believe their employees are ultimately responsible for their own well-being. Many of the academic theories on accidents focus on the worker or the environment in which they are working. In *Construction Safety*, Jimmie Hinze (1997) discusses the leading accident causation theories. Understanding these theories will help safety managers develop effective accident prevention plans.

Accident-Proneness Theory

The most common accident causation theory places responsibility firmly on the individual employee. Accident-proneness theory states that some persons have permanent characteristics that predispose them to a greater probability of being involved in an accident. This theory focuses on personal factors related to accident causation. It assumes that when people are placed in similar situations, some will be more likely than others to sustain an injury. Early researchers even suggested that accident proneness could be traced to personality traits (Vernon 1918).

While numerous studies have been conducted to determine the validity of accident-proneness theory, the research has mainly consisted of assessments of the distribution of injuries in a given population. Critics of this theory have argued that these studies neglected many variables that might contribute to accident causation. These studies often omitted variables such as personal problems, contributions of fellow workers, and occasionally differences in work hazards. Contemporary accident researchers generally accept that accident proneness may contribute to a small proportion of accidents, maybe 10 to 15 percent, but that there are many more variables that contribute to accident causation.

Newer research associates accident proneness with the propensity to take risks (Dahlback 1991). This study also suggested that workers with a higher propensity to take risks could be trained or motivated to make better choices.

Goals-Freedom-Alertness Theory

Goals-freedom-alertness theory states that safe work performance is the result of a psychologically rewarding work environment. Under this theory, accidents are viewed as low-quality work behavior occurring in an unrewarding psychological climate. This contributes to a lower level of alertness. A climate "richer" in diverse economic and noneconomic opportunities leads to a higher level of alertness and that alertness results in high-quality, accident-free work.

According to this theory, first suggested by Kerr (1950), a rewarding psychological climate is one where workers are encouraged to participate, set attainable goals, and choose methods to attain those goals. They must be allowed to participate in raising and solving problems. Goals-freedom-alertness theory essentially states that management should let workers have well-defined goals and the freedom to pursue those goals. The result is a higher level of alertness and a focus on the tasks at hand. This leads to higher quality work, safer behavior, and less accidents.

Goals-freedom-alertness theory suggests that managers and supervisors should try to make the work more rewarding for workers. They may use a variety of managerial techniques, including positive reinforcement, goal setting, participative management, and clear work assignments.

Adjustment-Stress Theory

Kerr (1957) proposed another accident causation theory to complement goals-freedom-alertness theory. Whereas goals-freedom-alertness theory maintains that a worker will be safe in a positive work environment, the adjustment-stress theory describes the conditions under which a worker would not be safe. Adjustment-stress theory states that safe performance is compromised by a climate that diverts the attention of workers.

This theory states that "unusual, negative, distracting stress" placed on workers increases their "liability to accident or other low quality behavior" (Kerr 1957). Similar to goals-freedom-alertness theory, the adjustment-stress theory focuses on the nature of the work climate as a major factor in accident occurrence. The climatic conditions impacting the worker may be either internal or external. Internal conditions include fatigue, alcohol consumption, loss of sleep, drugs, disease, or psychological stresses such as worry, personal problems, or anxiety. External conditions may include noise, illumination, temperature, or excessive physical strain. If the worker cannot adjust to the internal or external stresses, the chances for incurring an injury increase. In adjustment-stress theory, stress diverts the worker's attention during work hours and that diversion increases the susceptibility to injury.

Adjustment-stress theory suggests that managers and supervisors can actively work to alleviate stresses in the work environment. Managing both internal and external stresses decreases the probability of accidents.

Distractions Theory

Distractions theory states that safety is situational. Because mental distractions vary, the responses to them may have to differ to maintain safe performance.

Additionally, hazards, or physical conditions with inherent qualities that can cause harm to a person, may or may not be recognized by the worker and influence the safety of the task. Distractions theory was developed by Hinze (1997) to apply to a situation in which a recognized safety hazard or a mental distraction exists and there is a well-defined work task to perform. In the absence of hazards, there is little to prevent workers from completing their tasks. However, in the presence of hazards, work is greatly complicated.

Distractions theory has two components, the first dealing with hazards posed by unsafe physical conditions and the other dealing with a worker's preoccupation with issues not directly related to the task being performed. The theory basically states that when a worker has a higher focus on a hazard, the worker has a lower probability of injury and a higher level of task achievement. When a worker has a higher focus on a mental distraction, the worker has a higher probability of injury and a lower level of task achievement. To avoid injury and achieve high levels of productivity, workers must avoid mental distractions.

Chain of Events Theory

Sometimes accidents are the result of a series of events. When all of the events in the series occur, an accident results. If any one event in the series was omitted, the accident could have been avoided. This theory is called chain of events theory.

The last event preceding many worker injuries is usually an action performed by the injured worker. Often, the injury is blamed on worker behavior because the worker is simply the last party involved in the chain of events. However, stopping the occurrence of any event in the chain would break the chain and thereby prevent the accident.

To promote safety, it is important to consider all events in the chain, not just the final action of the injured worker. This theory recognizes that other parties play a role in influencing worker behavior. All parties associated with the various links in the chain of events have the opportunity to alter the course of events and thereby prevent accidents.

PROBLEM WITH ACADEMIC ACCIDENT THEORY

While these theories focus on the work environment or state of mind of the employee, it is still the responsibility of the employer to provide a safe working environment. Ultimately, safety is the responsibility of the employer. While most safety programs deal unwittingly with the causes of accidents proposed in the academic accident theories, such as eliminating unnecessary workplace hazards that complicate work, mitigating distractions such as noise that can prevent an employee from putting his or her full attention in the task at hand, and attempting to create a rewarding work environment to keep employees focused and attentive, few employers can successfully deal with the human element of an accident.

The Occupational Safety and Health Administration (OSHA), created under the Occupational Safety and Health Act of 1970, determined that employers such as contractors are responsible for providing workers a place of employment free from recognized hazards. This may be interpreted as a safe place at which to work. Accordingly, the manner in which a construction firm manages safety depends on the type of work in which the company is engaged.

A general contracting firm may employ hundreds of workers as employees, and many more as specialty contractors. The general contractor thus may have the responsibility of managing the safety of its own employees as well as the employees of the various specialty contractors utilized for the project. This often places the general contractor in an awkward position, since the firm may not be competent in managing the safety of all specialty contractors on the job-site. This is not meant to imply that the general contractor should assume the burden of implementing a safety program for every specialty contractor, but the general contractor should have a working knowledge of safety procedures for every specialty contractor on the site. This is often the most difficult aspect of implementing a safety program, because a thorough knowledge of safety practices is often learned on the job and usually requires considerable experience. As a result, the general contractor often leaves the responsibility of safety to the individual specialty contractors and may never take an active part in ensuring that the specialty contractor is actually exercising all measures necessary to provide a safe working environment.

Specialty contractors are faced with similar problems in that, like general contractors, they may employ large groups of workers, or as few as one. Like many other aspects of the work, the general contractor may attempt to pass responsibility for safety to the specialty contractor. Additionally, on projects with complex mechanical or electrical scopes of work, the specialty contractor may act as the prime contractor. Therefore, the primary problem faced by specialty contractors involves implementing a safety program that will satisfy the requirements of the general contractor as well as the standards outlined by OSHA. This may seem as an easy task; however, often the specialty contractors' emphasis on safety is proportionate to the size of the company. As an example, smaller companies may not place as high a priority on safety as larger companies. This, of course, is a generalization. While there are smaller firms with excellent safety programs and records, and while there is no doubt that smaller firms would benefit from a more comprehensive safety program, it is nonetheless a difficult process for them because of the expense incurred in implementing such a program.

Safety training is often left to an on-the-job learning exercise or taught by the employees' union or trade organization. While most union and trade organizations have excellent training programs, they should be considered only as a base upon which to build. The best training is often acquired through experience, on-the-job training, and continuing education.

The bottom line is that smart contractors, whether they are general contractors or third-tier specialty contractors, are assuming the responsibility for safety. The current economic conditions have determined that the financial liabilities associated with a poor safety record can make it impossible to turn a profit or even stay in business. Also, as has been previously discussed, poor safety performance may prevent a contractor from the opportunity to secure new work. Whereas the practice formerly was that only large general contractors employed full-time safety coordinators, now it is commonplace for trade specialty contractors of all sizes to employ someone whose primary duty revolves around the safety of its workforce. Whether formally or not, most companies have completed a rudimentary cost/benefit analysis of spending money up front on safety equipment, safety training, and safety programs to encourage workplace safety as opposed to saving that money and taking a reactionary response to workplace accidents. Most have likely determined that the money spent up front will more than pay for itself in the form of a lower EMR and insurance premiums, a decrease in the cost of its labor force, and a "safe" reputation in short order.

MAGNITUDE OF THE SAFETY ISSUE IN THE CONSTRUCTION INDUSTRY

Some trades are inherently more dangerous than others. A linesman working on overhead power distribution lines is more vulnerable to fall hazards and electrocution than a carpenter installing hollow metal doors and hardware in an office building. While the investment in safety required by the companies for each of these workers may differ, the fact is that all trades need to make that investment.

The dangers that face an electrician while on a jobsite, or potentially other trades while working in proximity to ongoing electrical work are high. Furthermore, they are not always as readily identifiable as other dangers associated with construction. For example, any worker who steps on a construction site should be able to clearly identify that the edge of a building 100 feet from the ground presents a safety risk if no handrail is present. However, it is not always clear whether a piece of wire laying on the ground or sticking out of a wall is live or not. Also, because of the nature of electricity, physical contact with an electric source is not required. Simply being too close to it with the wrong type of material can result in electric shock. Electrical contracting is a trade that requires a significant amount of employee training before stepping foot on the jobsite, as well as time spent learning about the hazards specific to that site. It is also important to make the investment in further employee training as predicated by job conditions, and to make the necessary investments in the equipment that can protect workers from accidents.

As with any trade, safety is not something that is solely the responsibility of those in the field to promote or maintain. Safety is a top-down cultural priority—that is, companies whose management places great importance on safety tend to carry that same view into the field, while a company whose management devalues safety for the betterment of the bottom line will likely have field employees who also share the same belief. If a management team conveys the importance of safety to employees by offering to pay for training, providing new safety equipment to replace old and worn equipment, and rewards workers for safety, their employees will be encouraged to work safely. Conversely, if that management team does not offer any employee safety training, punishes them for taking the time to properly use safety equipment at the expense of performing their work, and extols productivity over everything else, then their employees may take unnecessary risks or generally not create a safe workplace to make mandated productivity goals. While the management in the second example may meet their production goals, they do not take into account the fact that a single preventable accident may nullify any gains made by promoting production at the expense of safety.

Fatality Statistics

From 1997 to 2006, fatalities among all trades in the construction industry have ranged from 1107 total fatalities in 1997 to 1226 in 2006 according to the U.S. Bureau of Labor Statistics. In 2006, the construction industry accounted for the highest total number of workplace fatalities of any industry in the United States, and only agriculture with 29.6, mining with 27.8, and transportations and warehousing with 16.3 had more fatalities per 100,000 workers than the construction industry's 10.8 fatalities per 100,000 workers. The general trend over this 10-year period has been for an approximately one percent increase in fatalities per year. Figure 7.1 shows the number of fatalities per year for the period 1997-2006. The bottom line is that the construction industry is inherently dangerous, one of the most dangerous in the world. But it is somewhat promising to note that the increase in the number of man-hours spent working in the construction industry per year and the increase in the number of new trades people per year are both outpacing the rise in safety incidents. In fact, the total number of workplace related nonfatal injuries has decreased in the construction industry from 11 per 100 workers in 1997 to 6 per 100 workers in 2006, according to the U.S. Bureau of Labor Statistics. But even though this is the case, there is no such thing as an "acceptable level of losses." As more and more people join the industry, it is

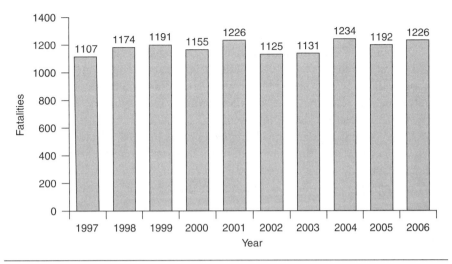

Figure 7.1 Yearly construction fatalities

incumbent on the construction companies, project owners, and design professionals to ensure that the emphasis placed on safety continues to grow.

Figure 7.2 shows the total number of fatalities for several trades within the construction industry in 2006. Laborers accounted for more than 27 percent of the total fatalities. However, laborers also account for the highest number of

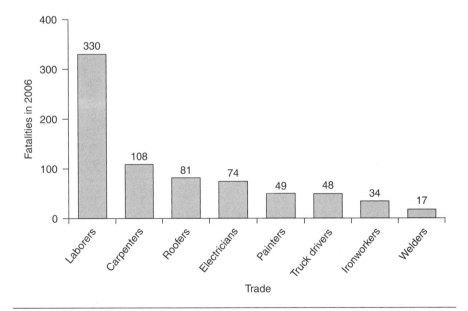

Figure 7.2 Construction fatalities by trade in 2006

tradesmen within the construction industry, which serves to mitigate the disparate number of deaths. Falls accounted for nearly 36 percent of the fatalities among laborers. Electrocution, either by contacting live equipment or overhead power lines was the cause of death among laborers in 7 percent of the cases. Among both carpenters and roofers, falls were a common cause of a workplace fatality. Death by electrocution was also common for carpenters, typically from coming into contact with live equipment.

Of the 74 fatal accidents involving electricians in 2006, 70 percent were the result of electrocution. More than half of these were the result of working on or near live electrical systems. The second-most common cause of fatality was falling, and most of these were due to a minor electrocution event preceding the fall.

METHODOLOGY

The primary objective of this study was twofold: first, to identify recommended safety practices and to compile information that could be put together so that individual companies could tailor to their specific needs; second, to perform a detailed cost/benefit analysis that would show quantitatively that the cost of implementing safety into each job a company performs is less than the costs incurred through unsafe practices in the form of accidents and injuries, damage to property, and an increased EMR.

The data for the analysis were obtained by mailing two separate detailed questionnaires to a selection of electrical contractors throughout the country. The hope was to include electrical contractors performing all types of work in the industry, including line work, new commercial construction, new residential construction, and maintenance and repair. The initial questionnaire was sent out to determine the interest of companies as potential participants, and to introduce the topic of study to each of the companies. It was requested that each company's safety director or other duly appointed employee answer some general questions, and submit the responses along with a copy of the company's existing safety manual. The second questionnaire was sent to each responsive company to develop an in-depth view as to the level of safety training, education, and orientation provided to the employees, and the standard safety practices that are employed on a typical jobsite in conjunction with and aside from the safety criteria required by the project owner and general contractor.

A total of 518 companies were contacted requesting participation in the study as well as the names of their safety directors. We received 125 responses (24 percent) with some contact names who have agreed to participate further in this study. Detailed questionnaires were then sent out. Out of those 125, 106 surveys were returned.

Key components of what an effective safety manual should contain were identified based on comparison with high-quality manuals in various trades, and through input from industry professionals. Each manual was evaluated as to whether it contained these significant sections, and what other, if any, special provisions each contained, specifically pertaining to that company's scope of work. The information contained in each manual and the analysis of each assumed that a company would likely need to add information that pertained to its specific scope of work.

The second questionnaire consisted of short-answer, fill-in-the-blank, and multiple-choice questions. The questions were designed to elicit qualitative answers that would show the participant's use of safety techniques and equipment. Some questions focused on the overall importance of safety in the company's management, while others were designed to determine specific practices utilized by field-workers. The questions were either kept general enough, or were worded in such a manner that took into account the differences in the scopes of work of the various companies. Specific data were collected from each company with regard to their EMR, number of man-hours billed in the most recent calendar year, number of OSHA-recordable incidents, number of man-days lost, and when possible, a monetary cost to the company of all incidents.

A model was created as an attempt to make an accounting for the total cost of accidents and injuries suffered on a jobsite, and the expense to savings ratio of implementing a highly structured safety program consisting of significant employee training, and the purchase of all of the necessary equipment to create a safe working environment. Many of these costs of an injury or implementing a safety program are easily quantified, such as medical expenses, lost wages, and damage to property. The same holds true for implementing a safety program since the cost of eye and ear protection, safety harnesses, and employee safety training classes can easily be determined. However, the far-reaching impacts of an injury, such as loss of productivity or morale suffered by workers on a jobsite following an injury, or the jobs a company loses due to a high EMR, are difficult to identify and on which to place a value, as is the added value of worker confidence in his or her safety, or the increased morale of an employee who knows that personal safety is of utmost importance to his or her employer.

ANECDOTAL EXPERIENCES: RECOMMENDED PRACTICES

Recommended practices in safety can be found throughout the construction industry. This section includes not only anecdotal experiences of electrical contractors, but also experiences gathered from safety managers associated with general contractors, construction managers (CMs), manufacturing facilities, and insurance providers during the course of this study.

These anecdotal experiences are roughly sorted into categories: effective communication, pre-planning programs, teaming programs, incentive and assistance programs, upper management, safety equipment, and enforcement. The sum of these experiences is the safety attitude and culture that should become the standard across the industry.

Effective Communication

Voice Mail as a Safety Tool

Many safety officers are busy attending pre-construction meetings and preparing risk control plans for projects. Oftentimes, people who call end up getting voice mail. One safety officer uses voice mail to reinforce the safety attitude by saying, "This is John Smith and I'm away from the phone so please leave me a message. And remember, it is unacceptable for anyone to get hurt on an ACME project—so do your part." Every day tools, including voice mail can be used to reinforce the safety attitude.

Valuing Employees and Their Work

When a company shows that it values its employees by providing fair wages, good working conditions, and a feeling of a secure future, it is indirectly promoting a safer work environment. Employees who understand that they are appreciated and needed on-site, healthy, every day are safer and more productive for the company. They recognize that their family and friends, who depended on them, including their employer, cannot afford for them *not* to work safely.

Use Mistakes to Make You Better

Most mistakes that cause accidents have been made many times before. The only way to avoid making the same mistake is to expose it, publicize it, and tell everyone how to avoid it. One of the best ways to share experiences is in the periodic safety meeting. When there is a safety problem on the job-tell everyone. And, be honest. If someone has messed up, everybody knows anyway. Honesty builds both credibility and trust.

A Picture Is Worth a Thousand Words

Tell it like it is. One safety professional shows graphic pictures of fatal accidents to communicate the shock and horror that overcomes a crew when a severe accident occurs. He has found that just talking about a situation is ineffective and some employees respond better to visual examples. However, the photos are always of accidents in other states because there may be members of the group involved (family, friends) with local tragedies.

EMR Fluctuation as a Communication Tool

Let people know that their jobs are at stake. One safety officer shows graphs with the EMR fluctuation history of a company (not identified, of course) that has had accidents against graphs of the financial history of the same company. He explains how failure to pay attention to safety can cause a company to fail. Impressing the importance of safety and showing the ramifications of unsafe acts on the financial strength of the company get employees' attention.

Continuously Reinforce Safety

Several safety officers have suggested that discussing safety only at weekly meetings is not enough to reduce the high number of losses in the industry. A fundamental change in the way contractors conduct their business on a day-to-day basis can significantly impact their potential for loss. To change people's way of thinking about loss and loss potential, some safety officers suggest talking to employees about safety each and every day before they begin their work. They claim that repetition is the key to change, and this cannot be accomplished if the only time the company talks about safety is during the weekly safety meeting.

Special Safety Symposium

One company has a special safety meeting every other week where they bring together everyone working on the jobsite for a sponsored lunch and "safety symposium." Anyone working on the job is included, from apprentices to management to the owner. The topic highlights a current activity on the job and reviews safe practices, regulations, and OSHA guidelines. The format is always different; sometimes they have a guest speaker and other times they watch a safety video.

One of the most productive meetings occurred when the general contractor showed the results of a job-site inspection checklist in which foremen from all the trades used digital cameras to document safe and unsafe working conditions on the jobsite. During the meeting, the photos were displayed on a screen. As each photo was shown, the person who took it described why they took the photo and everyone provided input. It is informative, educational, and even entertaining. The open discussion allows everyone to see the job from another perspective.

Do Not Hide Behind the Numbers

One company was enjoying a strong EMR near 0.7, but experienced an accident that made the CEO realize they had previously been content to look at the numbers and hide behind them. The company decided to change and involve the entire organization in safety. To do this, they decided to discuss safety at every meeting held within the corporation. The first item on every agenda, whether a Board of

Directors' meeting, a division review, or a business development seminar, became safety. What are you doing about it? They discuss every accident and every near accident at these meetings. As a result, their EMR now averages 0.5.

Do Not Try to Budget Safety

Many companies have come to the conclusion that they cannot and should not budget safety. They never question the cost of safety, only the outcome. Safety training programs are continuous. They are constantly updated, tailored, and refined as each project's needs change.

Safety Concerns for Minorities

There are many minority groups that are highly represented in the construction trades. The largest single minority group is Hispanics. Workplace accident fatalities among Hispanics exceed those of other groups by 20 percent, according to the U.S. Bureau of Labor Statistics. One reason is that Hispanics are disproportionately represented in dangerous and low-wage, low-skilled jobs where their safety training is often neglected. Adding to their special risk factors are language barriers, improper work documentation, and clashing social and cultural conditions. Two factors that impact the high Hispanic fatality rate are:

1. Assigning work before providing safety instruction. Hispanic hires are often sent to a jobsite without receiving minimal safety guidance. Formal training may be planned for a later time, but meanwhile, the new worker is thought to be ready for simple tasks considered "safe" on the assumption that every adult understands the U.S. work environment. Unfortunately, this knowledge is not universal among non-natives.

2. Failure to audit and review the success of safety instructions. Lack of follow-up to see the true level of comprehension probably plagues all safety efforts, but the need for diligence is doubly critical when there are language barriers. Double-checking and auditing must be factored into the safety training to ensure that the foreign-language workers grasp the concepts.

A worker who does not speak English as his or her primary language is at a much greater risk for injury simply due to the language barrier. This is not an issue that applies only to Hispanic workers. Any employer who chooses to take advantage of the willingness of a minority worker has an ongoing obligation to ensure that that employee has a full understanding of all of the risks around him or her, and the necessary safety measures that are to be taken with each task to be performed.

Pre-planning Programs

Safety and Health Pre-planning Is the Key

One safety officer has found that effective pre-planning is the key to success of job safety. He relayed the story of two craft employees who owe their lives to pre-planning: "A construction manager (CM) was managing construction of a concrete plant. A specialty contractor came on to the job to work on the plant's preheater tower, about 75 feet above the ground. The employees were to use a two-point power scaffold secured to the roof level. Per their safety program, the CM completed a task safety analysis. The specialty contractor told the CM's safety manager that independent lifelines were unnecessary because of the scaffold's 'new' design. A heated argument ensued, and ultimately the State's consulting safety professional agreed that independent lifelines were needed. That next morning, the scaffold's structure totally failed, dropping the employees until the newly installed lifelines saved their lives. The time spent to preplan safety will go far to eliminate the 24 fatality notifications delivered every week in the construction industry."

Want to Save a Life? Turn the Power Off

Electricians face a number of job-related hazards. Major injuries are caused by flash burns; electrical arc blasts causing molten metal burns, and electrocution. These hazards can be avoided if the power is turned off before beginning work on systems. There are few situations in which systems cannot be de-energized, but workers often run into resistance when they try to convince owners that shutting down is the best option.

One company uses a formal sign-off process that asks the owner to authorize the work on an energized system and details the possible hazards. The owner is made aware of the full range of risks. Once the full impact of a potential accident is understood, the owner often agrees to de-energize.

Overhead Power Lines Permitting to Prevent Electrocution

Similarly, one safety director recommends creating a permit system for opening an energized enclosure, such as a transformer or switchgear. The permit system should walk the worker through the lockout/tagout aspects of the operation. Additionally, the procedures identify the engineering, administrative, and personal protective equipment controls necessary to perform the work safely. The system can be structured similar to a confined space permit system to ensure that the proper precautions have been identified and confirmed prior to creating an unnecessary exposure to energized parts.

Production and Safety Working Together

Supervisors in the construction industry are constantly pressured (both internally and externally) by schedules and budgets. As a result, they may receive the message, either directly or indirectly, that safety is less important, and the job must go forward at any cost. To prevent this, safety and management personnel must visit sites regularly to show the project team that their safety efforts are supported and assistance will be available at any level needed to ensure a safe environment. One company owner suggested that during these visits, supervisors should be encouraged by upper management to take five minutes every day to talk with their crews about the hazards they may face during the coming workday and to discuss ways of minimizing the risks. This will show the field that all levels of management truly want everyone to go home safe to their families.

Time Out for Safety

One company uses a "Time Out for Safety" program that empowers individual workers who perceive there is a potential safety or health hazard, or lack of information associated with a task they are performing, to stop work and take a "time out" to address safety issues. During employee orientation, the program is introduced and management assures employees that they are guaranteed protection from retribution. Employees are given a wallet card signed by the project manager stating that it is their right and responsibility to stop work if there is a safety concern. Several projects even issue affirmative hard-hat decals to employees who have taken a "time out" and submitted their names for rewards.

Teaming Programs

Provide a Mentor for New Employees

A mentoring program provides many benefits to an employer. Most accidents, including fatalities, occur among new employees who have been on the job less than a year. Placing a new employee with a mentor (a veteran worker who is not their supervisor) has proven to increase the safety and quality of their work. It also reduces the amount of time it takes to teach them to work safely and efficiently. New employees tend to have fears that drive them to think they have to prove themselves. This program eliminates those fears and increases employee morale and overall job satisfaction, while reducing turnover. It also benefits the mentors because they are reinforcing their own knowledge while mentoring the new employee and are motivated to follow safe work procedures because they realize that they are setting an example.

Employee Focus Teams for Employee Buy-In

One company has found that the more involved employees are at the early stages of a decision, the greater the probability that the safety program will achieve the desired result. A highly effective technique of getting buy-in from employees involves the use of focus teams. Teams consist of a group of employees from a cross section of jobs within the organization. The process can be used to address major program concerns or individual issues. The company provided the following example: "We were experiencing a serious noise exposure problem on a project so we conducted a one day focus team meeting. The employees recommended a 100 percent mandatory hearing protection requirement. Compliance with this requirement has been outstanding due to the employee buy-in. There have been no noncompliance or employee complaints about the program due to the process used to create the solution."

Incentive and Assistance Programs

Employee Assistance Programs

Some employees who are dealing with problems at home cannot be entirely focused on the task they are performing. With the knowledge that some employees need help to deal with personal, marital, or substance abuse problems, some companies have implemented employee assistance programs. The program offers free, confidential counseling sessions to not only the employee, but also to their family. It is better to provide assistance to an employee than to fire the individual. The employees get help and are protected from losing their job. Dealing with emotional and substance abuse issues through an employee assistance program is a win-win for everyone. It provides the company the opportunity to help employees deal with their daily lives. Through this support there is a decrease in lost workdays, accidents, and most importantly injuries and fatalities.

Incentive Programs to Raise Awareness of Safety

Incentive programs can be an effective way of raising workers' awareness of safety issues. Many programs work well; however, the most effective offer incentives on a more frequent basis in a team atmosphere. On one jobsite for a general contractor who had been experiencing numerous accidents, a superintendent and project manager told their crew that if they went three calendar weeks without an accident or injury, the company would have a catered lunch delivered to the jobsite and served by a local restaurant.

The program was so successful that specialty contractors on the project asked to be included in it. They were offered the same incentive—no accidents or injuries over a three-week period and they would be included in the catered lunch, with the general contractor picking up the tab for the lunches. The result was a dramatic decrease in accidents and injuries. Consequently, the general contractor began using this program on all of its projects, with the overall result being less accidents and injuries.

Preoperational Workplace and Equipment Inspections

One company was particularly concerned about mobile equipment. With so many employees, it was not possible for supervisors to watch all the operators to ensure they inspect their equipment. The safety department devised a system to help supervisors determine whether inspections were being performed. Each supervisor was provided with a set of 5 × 7 inch bright yellow magnetic cards that read, "If you find this card, immediately contact your supervisor" in bold letters. The supervisors were directed to place the cards in areas that should be inspected by employees prior to shift start-up. If an employee contacts the supervisor and presents the card before start-up, they are rewarded with a $10 gift card. If the employee does not contact their supervisor within 30 minutes, the supervisor is to stop the vehicle, review the inspection sheet, show the operator the sign, discipline the employee, and explain the consequences of improper safety inspections.

Upper Management

Top Management Training to Instill Safety as a Corporate Value

At one company, all operations and senior management staff are required to have a current annual certificate of environmental, health, and safety training to continue employment. The training consists of a minimum of 8 hours for management and 16 hours for field staff. Every senior manager, including the owner, is required to visit a project to conduct a documented safety review at least once per quarter. Each review is then used to determine additional training needs for either the project team or the senior management group. This process has helped to instill safety as a corporate value and has had a profound effect on safety.

Safety Equipment

New Hire Striping

Several companies have implemented hard-hat striping programs to recognize new employees in the field. New employees are given a one-inch reflective red stripe that is worn front to back over the top of the company-issued hard hat for a six-month period. Apprentices are issued a one-inch reflective green stripe, which is worn side by side to the red stripe.

The purpose of the striping program is to allow for quick identification of skill base and knowledge, which facilitates safe production in the field. If, during an operation, there are a number of new employee hard hats together, the project team can quickly make certain that supervisory personnel are also in attendance. Moreover, it has strengthened communication among workers by helping to identify the experienced individuals from those who are new to the company. Incidentally, the reflective stripe(s) increase visibility for night work.

Automated External Defibrillator on the Jobsite

One company owner believes that all jobsites should have at least one automated external defibrillator (AED) on-site. He has a first responder's kit, oxygen, and an AED in his truck and in his safety engineer's truck at all times. The owner relayed the following explanation: "Several years back, I probably could have saved a 23-year-old man who had an irregular heartbeat if the kits had been on the market at the time. I got to the man in less than five minutes and started CPR while another man did rescue breathing. The emergency squad got there in about 12-15 minutes and hit him with the paddles. This got a heartbeat temporarily, but they lost it and he died. To this day, I believe I could have saved his life if an AED was available on-site."

Using Your Head in an Emergency

One safety team set up a program to make certain that an injured worker's family is contacted as quickly as possible. They require every on-site employee to maintain a "hard-hat emergency card" that lists a primary contact, phone number, language spoken, and any preexisting medical conditions the individual may have. A new card is reissued annually to ensure the information is current.

Workers place the card in a plastic adhesive sleeve inside their hard hat. The card saves precious time during an accident because personnel do not have to dig through the employee's file to find basic information. Because no worker is allowed on-site without his or her hard hat, it is the best place for the emergency card. Since calm heads do not typically prevail in an emergency situation, they have printed CPR instructions on the back of the card for quick reference.

Enforcement

Occasionally Switch Safety Inspectors

One safety team suggests switching safety inspection duties regularly to stay sharp. Inspectors get so accustomed to seeing the same thing every day that it becomes easy to overlook small details. When it comes to safety, small details can make the difference between a job well done and a fatality. It is often a good idea to have a "set of fresh eyes" occasionally walk the site.

SAFETY PROGRAM ANALYSIS

Table 7.1 was developed based on the results of the questionnaires that were sent to participating contractors as explained in the Methodology section. Each company was asked to give their opinion about various components of what would be considered a "typical" safety program. Respondents were asked to give a ranking on a scale of 0-5 of their perceived value of each component. A zero indicates that a particular component is not utilized; a one indicates little effectiveness; a five indicates the highest level of effectiveness. Table 7.1 shows the average values for each category. Respondents were then asked to provide information as to the cost of each component. Using the average number of people that each company had under employment during an average year, an average cost per person for each component was determined. The relative value of each component was determined through a simple cost/benefit analysis by dividing the annual cost per person per year of each component by the average perceived level of effectiveness as a means of comparison of the components.

Components of Effective Safety Programs

The components of an effective safety program can vary widely depending on the size of a given company and the type of work that it typically performs. To standardize the questionnaire, the components were kept as generic as possible. Components ranged from a full-time safety director, to a safety incentive program, to daily stretching prior to beginning work. Respondents were also given the opportunity to provide their input on other components that should be considered in preparing a company safety program.One item that was mentioned was a jobsite weekly safety meeting. Many of the items in the questionnaire would qualify as potential safety meeting topics.

Relative Effectiveness of Components

The relative effectiveness of each component of a safety plan again will vary greatly depending on the size of a company and the type of work it typically performs. However, the components listed in the questionnaire can be modified as necessary to provide a company-specific safety plan.

Pre-task planning, prompt and thorough incident investigation, a full-time safety director, and proper lockout/tagout procedures were determined to be the most effective components of a safety plan. With the exception of a full-time safety director, all of these would apply to each and every electrical contractor, regardless of size or scope of work. As has been previously noted, more than 50 percent of all electrocutions are the result of working on or near energized systems. Pre-task planning and proper lockout/tagout procedures may eliminate

Table 7.1 Components of a typical safety program ranked by effectiveness

Safety element	Effectiveness	Annual cost ($)	Cost/Person ($)	Cost/Benefit
Pre-task planning	5.00	15,000	47.16	9.43
Incident investigation	5.00	25,000	73.50	14.70
Full-time safety director	4.85	75,000	266.52	54.95
Lockout/Tagout procedure program	4.80	1,500	4.72	0.98
Fall protection work plan	4.50	1,000	3.14	0.70
Basic safety rules	4.40	1,000	3.14	0.71
First aid	4.20	500	1.57	0.37
Confined space entry program	4.20	5,000	15.72	3.74
Ladder safety program	4.00	2,000	6.28	1.57
Near miss program	4.00	2,000	6.28	1.57
Weekly safety meetings	4.00	300,000	650.00	162.50
Proper lifting techniques	3.70	1,000	3.14	0.85
Job hazard analysis	3.80	2,500	7.85	2.07
Drug testing program	3.60	18,000	41.57	11.55
Incentive programs	3.50	46,250	114.31	32.66
Hazard communication program	3.40	5,000	14.70	4.32
Requirements for power actuated tools	3.40	2,000	6.28	1.85
Hearing conservation program	3.00	500	1.57	0.52
Respiratory protection program	2.90	2,000	6.29	2.17
Project safety indoctrination sheet	2.30	3,000	9.42	4.10
Contractor safety questionnaire	2.20	1,000	3.14	1.43
Asbestos program	2.10	7,700	125.63	59.82
Blood-borne pathogens	1.90	200	0.63	0.33
Lead program	1.40	500	1.57	1.12
Stretch and flex program	1.10	5,000	14.70	13.36
Workplace violence program	0.50	500	1.57	3.14

many of these incidents, while prompt and thorough investigation can help to ensure that a similar incident is not repeated. While a full-time employee whose sole responsibility is safety may not be feasible for smaller companies due to the high cost or the lack of work required to justify a full-time safety officer, employing someone who spends 50 percent of his or her time as the safety director while also performing other functions is also an option that may be considered. Additionally, smaller companies may opt to require their key field people to receive added safety training and have each serve as the safety officer on his or her site.

Each of the components listed in Table 7.1 that is ranked a four or higher should be considered for inclusion in any safety manual or training program. All of them can be easily modified and adapted to fit a particular company's field of emphasis.

Annual Costs and Cost/Person

The rank order of the 16 least expensive elements shown in Table 7.1 does not change whether examined by annual cost or cost per person. There does appear to be a slight variation in the order of the more expensive elements. It may be of special interest to the smaller contracting firm to be able to identify those accident prevention programs that have an annual cost per person of less than $10. Table 7.2 shows those elements ranked from the least expensive to the most expensive.

Basic Cost/Benefit Analysis

The cost/benefit analysis in this study serves to show how truly inexpensive many of the most effective safety program components really are. Fifty percent of all prevention programs examined in this study had an annual cost per person of $6.28 or less. Most of us would conclude that the majority of the components with an average "effectiveness" ranking of four or higher have an annual cost per person that is less than its perceived value. That perception was supported by this study as the actual average annual cost per person for the 11 most effective programs was $98.

The cost/benefit is determined by dividing the annual cost/person by the effectiveness rating to see which accident prevention program gives a company the most "bang for the buck." Table 7.3 shows how the different elements rank using this cost/benefit measure. The smaller the value in the cost/benefit column the most return we can get for each dollar spent.

If we examine all the accident prevention measures in this study, we notice that many of the lower cost programs return good benefits for the amount spent. Many of the more effective programs are also inexpensive. It is also surprising that because of their high costs, some of the most effective programs appear low on the

Table 7.2 Components of a safety program with an annual cost per person of less than $10 ranked by cost

Safety elements	Effectiveness	Annual cost ($)	Cost/Person ($)	Cost/Benefit
Bloodborne pathogens	1.90	200	0.63	0.33
First aid	4.20	500	1.57	0.37
Hearing conservation program	3.00	500	1.57	0.52
Lead program	1.40	500	1.57	1.12
Workplace violence program	0.50	500	1.57	3.14
Fall protection work plan	4.50	1,000	3.14	0.70
Basic safety rules	4.40	1,000	3.14	0.71
Proper lifting techniques	3.70	1,000	3.14	0.85
Contractor safety questionnaire	2.20	1,000	3.14	1.43
Lockout/Tagout procedure program	4.80	1,500	4.72	0.98
Ladder safety program	4.00	2,000	6.28	1.57
Near miss program	4.00	2,000	6.28	1.57
Requirements for power actuated tools	3.40	2,000	6.28	1.85
Respiratory protection program	2.90	2,000	6.29	2.17
Job hazard analysis	3.80	2,500	7.85	2.07
Project safety indoctrination sheet	2.30	3,000	9.42	4.10

list. The bottom line is that if you have limited resources, a company should look to invest in the programs that give them the most return on their investment.

Recommended Practices Analysis

The results of the safety questionnaire provide insight as to several components that should be part of a quality safety program. Many of the components are applicable to all contractors, regardless of company size or scope of work. It is clear that when making decisions on where to invest your safety dollars, you must both consider cost/benefit and effectiveness together. For example, having a weekly safety meeting that all employees know about and attend on every jobsite is a safety measure that holds true for residential work, light commercial, and commercial

Table 7.3 Components of a safety program ranked by cost/benefit

Safety elements	Effectiveness	Annual cost ($)	Cost/Person ($)	Cost/Benefit
Bloodborne pathogens	1.90	200	0.63	0.33
First aid	4.20	500	1.57	0.37
Hearing conservation program	3.00	500	1.57	0.52
Fall protection work plan	4.50	1,000	3.14	0.70
Basic safety rules	4.40	1,000	3.14	0.71
Proper lifting techniques	3.70	1,000	3.14	0.85
Lockout/tagout procedure program	4.80	1,500	4.72	0.98
Lead program	1.40	500	1.57	1.12
Contractor safety questionnaire	2.20	1,000	3.14	1.43
Requirements for power actuated tools	3.40	2,000	6.28	1.85
Near miss program	4.00	2,000	6.28	1.57
Ladder safety program	4.00	2,000	6.28	1.57
Job hazard analysis	3.80	2,500	7.85	2.07
Respiratory protection program	2.90	2,000	6.29	2.17
Workplace violence program	0.50	500	1.57	3.14
Project safety indoctrination sheet	2.30	3,000	9.42	4.10
Hazard communication program	3.40	5,000	14.70	4.32
Pretask planning	5.00	15,000	47.16	9.43
Drug testing program	3.60	18,000	41.57	11.55
Stretch and flex program	1.10	5,000	14.70	13.36
Incident investigation	5.00	25,000	73.50	14.70
Incentive programs	3.50	46,250	114.31	32.66
Full-time safety director	4.85	75,000	266.52	54.95
Asbestos program	2.10	7,700	125.63	59.82
Weekly safety meetings	4.00	300,000	650.00	162.50

work. While you may want to maximize the benefits of the dollars spent on safety (cost/benefit), and while this program does not rank favorably from a cost/benefit standpoint as compared to the other programs, nonetheless, this program cannot be overlooked because it ranks as one of the most effective accident prevention programs. Similarly, any type of contractor can receive benefit from pre-task planning prior to beginning a new project or a new portion of a project. Items that fall under the heading of near-universal application are weekly safety meetings with all crew members to discuss specific topics, including incident investigation, lockout/tagout procedure program, fall protection, work plan and ladder safety; preparing a job hazard analysis plan prior to each project with which every employee must familiarize himself; company-wide drug testing; and a safety incentive program.

Other components may be specific to companies of a certain size, or that typically perform certain types of work. For example, a full-time safety coordinator would cost a company a significant amount of money per year. Only companies that are large enough to absorb this overhead cost, or perform enough projects where a safety director is a billable party, will see this as a feasible measure. This does not, however, preclude smaller companies from hiring an employee who acts as a part-time safety director, or retraining a current employee to act as a safety expert. Education on lead poisoning, asbestos poisoning, and proper respirator use are beneficial to any contractor, but again, they are most applicable to certain types. Companies that do significant amounts of remodel or tenant improvement work would want to ensure that their safety training includes these topics.

Prompt and thorough incident investigation is a portion of a safety plan that is accepted as crucial, and rightly so. Knowing how an accident happened is one of the easiest ways to prevent the same accident from happening again, can teach companies what the most dangerous aspects of their work are, and can highlight areas that require more employee training and education to prevent similar situations from ever developing. One safety measure that is nearly always neglected or even considered is to investigate near misses as opposed to just investigating actual accidents. The reason being is that the number of near misses is going to be greater than the number of accidents, and identifying the root source of a potential problem can keep it from ever becoming a real problem. The level of detail of the investigation can be as simple as the project foreman asking the parties involved with the near miss and any eyewitnesses what events led to the near miss and documenting the event in a journal. Undoubtedly, there is a cost associated with consistently investigating incidents where nothing really happened. However, when compared to the total direct and indirect costs associated with an accident, the 15 minutes someone might spend asking a few questions, plus the 15 minutes documenting the incident, plus 2 hours per month by a project manager or superintendent spent reviewing all of the near misses certainly falls well short by several orders of magnitude.

CONCLUSIONS

Further work needs to be completed to truly define a set of Best Safety Practices. And the wide variety of work is so broad and diverse that there probably needs to be a set of Best Safety Practices for each type of work.

Certainly further research needs to be done to determine the cost versus benefit of different aspects of a safety program. To do this requires that contractors provide detailed information regarding their expenditures on both direct and indirect costs associated with an accident, as well as more detailed information about their safety budgets. This level of detail was beyond the scope of this study, but would be useful in showing the bottom line.

Many of the respondents to the survey relayed the same message: that safety begins at the top and needs to permeate through every aspect of the company and become embedded in the corporate culture. They claim that the most effective method to create safe project sites and prevent accidents is to develop a safety culture that is ingrained in every individual. That culture should demand zero tolerance for unsafe acts. Training, education, and enforcement on a continuous basis develop this culture. The involvement of every individual (new hires, specialty contractors, and most importantly, upper management) is critical to achieving this attitude and culture.

Employees tend to respond better to training and education than they do when penalized for their actions. Once employees understand that the ultimate goal is to keep them safe, healthy, and employed, they quickly become advocates of safe practices. Additionally, they actively participate in the continuous monitoring for safety.

Safety training is the beginning of a long process of safety education. Once the rules are known, the next step is to establish a level of tolerance. Whenever a rule is not followed, either intentionally or otherwise, there is an opportunity to reinforce the safety attitude. Every violation of a rule needs a response. Most responses can fall along the lines of education and reinforcement. Occasionally a violation may require disciplinary action. Rarely is termination required if an employee is cooperative and willing to modify his or her safety attitude.

A positive safety attitude is infectious. Recognize those who are doing the job correctly, and everyone will try to follow. Belittle those who are doing something incorrect, and production will suffer—right along with morale.

Construction projects can take years of planning and preparation. Countless hours are spent utilizing resources to minimize the probability of problems, yet the industry still focuses on safety glasses and hard hats and slogans like "safety first!" It's not that glasses and slogans are wrong. However, successful safety programs require an attitude of responsibility that must be ingrained in all project leaders. This attitude must shout that it is morally and financially unacceptable for anyone to get hurt on a project.

WEB ADDED VALUE

SeguridadOcupacional.pdf

This file is a Spanish translation of this chapter. Given the increased number of Hispanics who join the construction industry every year, a Spanish version of this chapter may be helpful to contractors who may want to reach this audience in their own language.

8

TOTAL QUALITY MANAGEMENT

Dr. Thomas E. Glavinich, *University of Kansas*
Dr. James E. Rowings, *Kiewit Corporation*

INTRODUCTION

Total Quality Management (TQM) is an operational philosophy. This means that TQM is at the same time a management philosophy and a set of techniques. The term is described in different ways by different people. This variety comes about because TQM represents a combination of several fundamental concepts. The variations are due to individual preferences for the point of emphasis for these concepts. Generally, these many definitions of TQM will include several of the same key concepts.

The first key concept is the focus on customer satisfaction. Contractors cannot exist without customers. Clearly, to meet customers' needs a process must be in place to identify the customers' needs and expectations. To satisfy customers, contractors must not only meet all technical specifications, but also be perceived as meeting all the other nontechnical expectations of customers. As will be discussed later, there are both internal and external customers. This leads to a major effort in understanding who all your customers are and then identifying their needs and expectations.

A second key concept behind TQM is that of focusing on prevention of problems rather than on finding and correcting problems after they occur. The concept here is that it is better to design the processes so that effort is directed to doing things correctly the first time rather than in performing excessive inspection and rework. Processes must be carefully scrutinized and analyzed to identify the critical operations and activities that lead to quality, and to train and retrain the individuals performing the work to achieve the desired results the first time and every time.

A third key concept involves participation of all those involved in the processes. This involvement is necessary to expand the knowledge applied to the planning, problem-solving, and control efforts. This additional involvement in these processes achieves the desired goal of obtaining commitment and accountability for the results. The people performing the work are the best sources of knowledge for improving the processes. The responsibility for quality becomes jointly and individually shared by everyone through application of this concept.

TQM has at its heart the concept of continuous improvement. The TQM approach is one where the business is constantly and forever striving to improve beyond the current levels of performance. To achieve this goal, there must be measures of performance. TQM relies heavily on the systematic collection of data for decision-making and the distribution of the data and information to all the participants involved in the processes. This concept of continuous improvement is critical to long-term competitiveness.

FUNDAMENTALS OF TOTAL QUALITY MANAGEMENT

If TQM is practiced as outlined in this chapter, contractors will begin to realize several benefits. The customer will start to benefit and the process is one that will lead to repeat business. Performing satisfactory work and meeting customer expectations will also provide a base of customers that will provide referrals and strong recommendations to other potential clients. The close focus on customers' expectations and communication with customers can build a commitment on the part of repeat customers to your welfare. At a minimum there is a relationship of understanding and fairness that is established. The synergistic relationship that develops with the TQM approach builds mutual trust and creates a mutual dependency that can lead to greater business opportunities.

Implementing TQM in contracting firms should lead to faster jobs; less labor, materials, and equipment; less overhead expenses; less idle time; less rework; and less accidents. Contractors will streamline processes and make better use of human resources. Examination of production processes can reduce waste and inefficiencies thus reducing the labor, materials, and equipment required to complete projects. As processes are examined and modified, it is possible to reduce the idle time associated with the many interfaces with other trades. Focusing on doing things correctly the first time will lead to methods that will greatly reduce the amount of rework. Better planning will also reduce the number of accidents by selecting appropriate methods that have less opportunity for serious injury. These improvements all represent improved productivity for contractors.

Increased productivity will result in increased profit margins. A major portion of the savings received from the productivity increases will flow directly to the bottom line in the form of profit. Since fewer resources will be used to produce the same work, the return on the contractor's investment will be improved. The risk that contractors face will be reduced by having more complete planning and the commitment from the project team to make the methods and processes work. Continuous improvement involves taking a leadership position in the marketplace. Leadership in the marketplace allows for higher profits and means that fewer competitors are able to satisfy the customer the same way. This competitive advantage provides the opportunity for a higher markup.

By satisfying and exceeding customer's expectations, contractors position themselves to become the supplier of choice among more customers and thus increase market share. Customers will gravitate toward quality contractors. The best contractors over time will grow to meet the greater markets.

The focus on problem prevention reduces the need for corrective processes. Claims and litigation represent corrective processes that can be avoided by better communication earlier to establish understanding and trust in working relationships. By eliminating claims, contractors avoid the expense of pursuing a claim through arbitration or litigation. The time saved in avoiding these corrective processes can be applied to further improving other productive processes and furthering the competitive advantage.

The principles behind TQM have been around a long time and have been practiced by many organizations that were successful throughout the ages. TQM, like gravity, has always existed, but the leaders associated with it were those that recognized what it was that made some organizations successful, and developed a structure to explain it and a procedure for consistently practicing it. The most prominent leaders are Deming, Juran, and Crosby along with others who developed many of the tools employed in the TQM process.

Deming

W. Edwards Deming is by far the best known of the founders of this management movement. Dr. Deming began his work in statistical quality control during World War II and became the pioneer of the quality movement in Japan after 1950. Deming's approach to quality arises from his statistical background and his own philosophy of management. His philosophy can best be summarized with his "Fourteen Points" (Walton 1986):

1. Create constancy of purpose
2. Adopt the new philosophy
3. Cease dependence on mass inspection
4. End the practice of awarding business on price tag alone
5. Improve constantly and forever the system of production and service
6. Institute training
7. Institute leadership
8. Drive out fear
9. Break down barriers between staff areas
10. Eliminate slogans, exhortations, and targets for the workforce
11. Eliminate numerical quotas
12. Remove barriers to pride in workmanship
13. Institute a vigorous program of education and retraining
14. Take action to accomplish the transformation

In Deming's mind, the pursuit of quality represents a never-ending process. All of Deming's philosophy is embedded in his 14 points. Deming's 14 points and their applicability to the contracting industry will be discussed in a later section of this chapter.

Juran

Joseph Juran, like Deming, was actively developing his philosophy on quality in Japan in the post-World War II era. Juran also expanded from the area of statistical quality control to the whole area of management. Juran focuses on quality improvement at the project level. Juran has drawn analogies between finance and quality. Therefore, he equates quality control to financial control, planning to financial planning, and budgeting and quality improvement to cost reduction. Juran's contention is that if the same level of discipline and structure were applied to quality as is already applied to budgeting, real gains and improvements could be made. Juran has proposed the following trilogy as a solution (Juran 1988):

1. Quality Planning
   - Establishing quality goals
   - Identifying the customers, which are those impacted by the process used to meet the goals
   - Determining the customers' needs
   - Developing features that respond to customers' needs

- Developing processes that can deliver the features
- Establishing controls on the process
- Transferring the plans to the operating personnel
2. Quality Control
   - Evaluating actual quality performance
   - Comparing actual performance to quality goals
   - Acting on the deviation
3. Quality Improvement
   - Establishing the infrastructure needed to secure annual quality improvement
   - Identifying the specific areas or projects for improvement
   - Establishing a project improvement team with the responsibility for executing each improvement project successfully
   - Acquiring the resources, instilling the motivation, and providing the training for the teams to diagnose the causes, discover the potential remedies, and establish controls to maintain the improvement

Juran's definition of quality is "fitness for use." He measures the cost of poor quality as those costs that would disappear if products and processes were perfect. He suggests the use of Pareto analysis to identify the best improvement projects, the "vital few" from the "useful many." He asserts that one can plan a strategy only after the goal has been established. Juran requires setting project quality goals first and then developing the strategy.

Crosby

Crosby's background is somewhat different since he does not come from a strict statistical process control (SPC) background, but rather from a management focus. Crosby starts by asserting that quality is not just goodness or luxury. He contends that certain myths exist about quality that must be exposed. Quality is not intangible. Quality is not something that originates with the workers or exists only in the quality control group.

Crosby sets the goal for TQM as zero defects. Crosby is most famous for his assertion that quality is free since the costs are less than the benefits of perfection. Crosby's (1979)TQM philosophy is captured in his "Four Absolutes":

1. The definition of quality is the conformance to the requirements, not goodness. Management has three basic tasks to perform: (1) establish the requirements that employees must meet; (2) supply the materials that the employees need to meet those requirements;

and (3) spend time encouraging and helping the employees to meet those requirements.

2. Quality is achieved through prevention, not appraisal: Appraisal, whether it is called checking, inspection, testing, or some other name, is always completed after the fact. If these actions are used for acceptance, then the result is sorting the good from the bad. Each action produces a little pile of material or paper that has to be further evaluated. Appraisal is an expensive and unreliable way of getting quality. Checking, sorting, and evaluating only sift what is finished. What has to happen is prevention. The error that does not exist cannot be missed. The secret of prevention is to look at the process and identify opportunities for error. These opportunities for error can be controlled.

3. The performance standard must be zero defects, not "that's close enough." A company with millions of individual actions cannot afford to have a percent or two go astray. Less than complete compliance with the requirements of a performance standard could cause that to happen. Conventional wisdom says that error is inevitable. As long as the performance standard requires it, then this self-fulfilling prophecy will come true.

4. The measurement of quality is the price of non-conformance, not indexes: The best measurement for quality is the same as for any other, money. The cost of quality can be divided into two areas, the price of non-conformance (PONC) and the price of conformance (POC). Prices of non-conformance are all the expenses involved in doing things incorrectly. For service organizations, these costs include redoing work, fighting fires from errors, budget overruns, and others that are typically 35 percent of all operating costs. The price of conformance is what is necessary to spend to make things come out right. This includes most of the professional quality functions, all prevention efforts, and quality education. It also covers such areas as procedural or product qualification. It usually represents about 3 to 4 percent of sales in a well-run company. The price of non-conformance can be used both to track whether the company is improving, and as a basis for finding the most lucrative corrective action opportunities.

Crosby has developed a structured approach. His 14 stages, not necessarily executed in strict sequence because education may come before management commitment, are (Crosby 1979):

1. Management commitment
2. Quality improvement team

3. Quality measurements
4. Cost of quality evaluation
5. Quality awareness
6. Corrective action
7. Establish an ad hoc committee for the zero defects program
8. Supervisor training
9. Zero defects day
10. Goal setting
11. Error cause removal
12. Recognition
13. Quality councils
14. Do it over again

Crosby stresses the importance that pilot projects have on convincing skeptics. His philosophy includes getting everyone involved eventually.

WHAT IS QUALITY?

Defining quality as it applies to a particular business endeavor is the first step in the development of a TQM program for an organization. Quality, as it applies to the contractor, can be defined as follows:

Quality is providing internal and external customers with products and services that meet or exceed their expectations and needs the first time and every time.

There are several important concepts embedded in this definition of quality that should be addressed. These concepts include: (1) internal and external customers; (2) products and services; (3) customer expectations and needs; and (4) first time and every time.

First, every job or task can be thought of as a process in which inputs are transformed into outputs. Therefore, the person or crew performing the job or task has customers, which are the users of the process output. These customers can be either internal or external. Internal customers are people within the contracting firm who use what others in the firm provide. For example, the crew in the field would represent an internal customer to the contractor's warehouse crew for the delivery of materials to be incorporated into the project. External customers are people or firms outside the contractor's organization that use a product or service the contractor provides. Obvious external customers for a specialty contractor are the general contractor and ultimately the owner.

Second, construction is a service industry where contractors bring the necessary labor, material, and production equipment together to complete the work defined by the contract documents. Contractors use their knowledge and expertise to plan, organize, monitor, and control the construction process to ensure that the work is completed on time, within budget, and in accordance with the plans and specifications. The contractor is ultimately responsible for the quality of the work. In addition to providing a service, some contractors also produce products that are integrated into the construction project. These products can include such things as custom control cabinets fabricated off-site in the contractor's shop. Similarly, the product could be a program for a programmable logic controller that will be used to control an industrial process. Again, no matter what the nature of the product, quality must be built into that product by the contractor.

Third, TQM is an operational philosophy that focuses the firm's efforts on achieving total customer satisfaction and the continuous improvement of its processes. With TQM, customer satisfaction is paramount. However, in construction it is important to recognize that the project delivery system employed and the contract award process for a particular project places constraints on contractors with regard to quality. Customer requirements are a function of the project delivery system employed. In negotiated work, contractors have a much greater ability to bring their unique knowledge and expertise to bear on the customer's needs and provide systems that are safe, functional, reliable, and economical over their intended life. However, on the other end of the spectrum with lump-sum competitive bidding, quality must be defined by the contract documents. The contract documents include the plans and specifications, which are normally prepared without input from the contractor. This process limits the contractor's ability to affect the quality of the installation except through the submission of voluntary alternates, which may or may not be considered. The need to obtain work on an ongoing basis and make a reasonable profit to continue as a going concern places constraints on the contractor regarding the definition of quality on lump-sum competitively bid projects.

Finally, the goal of TQM is to do the job correctly the first time and every time. Avoiding rework results in both increased productivity and improved morale.

Now that quality has been defined, the next step is to begin the implementation process. This implementation process must ensure that all the key elements necessary for an effective TQM program are addressed and incorporated into the contractor's continuous improvement process. To facilitate implementation of TQM within the contracting firm, the implementation process has been broken down into the following seven steps:

1. Committing to quality
2. Planning for quality

3. Organizing for quality
4. Managing for quality
5. Identifying improvement opportunities
6. Implementing TQM in the field
7. Measuring results

These steps are covered in the next seven sections of this chapter.

COMMITTING TO QUALITY

Committing to quality requires both leadership and vision on the part of contractors. Both of these attributes evolve into a formal mission statement for the firm and ultimately a strategic plan to guide the firm in the accomplishment of its mission.

When formulating the mission statement and strategic plan, management is committing the firm to a particular future. The philosophy of continuous improvement must be embedded in the corporate culture, which means that quality must be an integral part of the contracting firm's mission statement and strategic plan.

Leadership

Management formulates policies and implements strategy to achieve the firm's goals and objectives. Management, however, cannot merely direct. Management must lead if the firm's goals and objectives are to be achieved.

To be successful in today's business environment, management must be able to mobilize, excite, and focus the firm's energy on the accomplishment of strategic goals and objectives. To achieve this, management must supply the organization with a clear vision of where the firm is going and how it is going to get there. This is especially true for the firm that has adopted participative management as part of its TQM program where employees take greater responsibility for day-to-day operations. With participative management, employees need a beacon to guide them and ensure that their day-to-day decisions are consistent with the firm's long-term direction.

It is management's responsibility to provide the beacon for employees through the firm's mission statement. The firm's mission statement embodies management's vision of the firm's future. From the mission statement, specific goals and objectives can be identified that need to be accomplished within the planning horizon. To achieve these goals and objectives, a strategic plan should be developed to guide the firm toward its stated mission.

Vision

When you formed or took charge of your contracting firm, you had a vision of what could be. This vision may have been as simple as having a "successful contracting business." Vision is the creative thinking that you brought to the firm. Vision is what motivated you to form or take charge of your company and vision is what keeps you in business each year.

Vision is not about problem solving. Vision is about the future. With vision you leap through time to see the way you want things to be, not how they are today. Management must project a vision of the firm's future that both activates and motivates employees. Vision is the basis for the formulation of the mission statement and eventually for the development of the firm's strategic plan.

Vision also provides corporate identity. Vision unifies the organization from within and has a direct impact on corporate culture. Common vision is essential to the contracting firm and this common vision must be embodied in a mission statement.

Before you proceed with the development of a mission statement, you need to revisit your vision. Over the years, it is not unusual for vision to get lost in the day-to-day struggles of running the firm. You must take time to reaffirm or revise your vision of your firm's future before proceeding with the development of a mission statement.

Mission Statement

Based on your vision, the next step in the process is to formulate your firm's mission statement. As noted, your mission statement will embody your vision for the firm and be the cornerstone of your strategic plan. Your mission statement should answer the basic question, "Why is my company in business?" The mission statement is the catalyst that keeps the organization moving toward the realization of your vision.

The mission statement sets out the overall purpose of the organization. Your contracting firm exists to accomplish something in the larger business environment. The development of a mission statement for your company provides guidance to you and your employees when faced with alternative courses of action or tough business decisions. The mission statement should evolve from the following two basic questions: What is our business? And what should our business be?

The first question deals with what is going on today within the business and is important for establishing a baseline for the second more important question. Defining what your business should be in the future is what this whole planning process is all about. It is the response to the second question that eventually evolves into your mission statement. Your vision of the firm's future must be translated into a formal mission statement that can be communicated throughout the company and used as the basis for strategic planning.

The mission statement for a contracting firm is a statement of purpose that galvanizes the firm's employees with a sense of direction and opportunity. The formal mission statement shares your vision of the future with everyone in the company. Without a formal mission statement, there is no way that your vision of the firm's future can be communicated to the firm's employees. Without a unifying vision, the firm adopting participative management will drift because employees will not have the guidance they need to make consistent day-to-day decisions.

The owner or upper management of the contracting firm needs to formulate the mission statement. This mission statement needs to consider customers, market, social, economic, other external factors, as well as internal factors. The mission statement needs to address the firm's business and markets as well as its concept of itself and its culture. Elements of the mission statement can sometimes be found in the corporate charter or other corporate documents such as marketing literature.

Strategic Planning

The mission statement sets the direction for the contracting firm. The next step is to plan how the firm will work toward fulfilling its mission. There is usually a gap between vision and reality. If you have formulated an effective mission statement, that mission can never be achieved, only approached. This is the essence of TQM and the philosophy of continuous improvement. The mission statement provides the ideals to which the firm aspires. The journey toward fulfillment of the mission statement is never ending and is the basis for continual improvement of the contractor's services.

Formulating a strategic plan is the action that management takes today to achieve the firm's vision tomorrow. The strategic plan is the mechanism that the firm uses to work toward the realization of its mission. The strategic plan relates the contracting firm's internal situation to a changing external environment.

The strategic plan provides concrete goals and objectives that motivate employees and provide a method to measure progress toward achieving the firm's stated mission. Strategic planning is deciding where you want to go and how you are going to get there. Strategic planning is aimed at accomplishing the corporate mission.

Strategic planning differs from operations planning, which focuses on the day-to-day activities of the firm. Operations planning focuses on the scheduling of tasks that need to be accomplished to keep the firm viable in the short run.

You may not have a formal written strategic plan, but chances are you have a mental plan of how you intend to achieve your vision of where your contracting firm is going in the future. In most cases, the owners of successful contracting firms carry their strategic plans around in their heads and not formally written down. In the traditional hierarchical management system, this may be acceptable since the only person who needs to know where the firm is going and makes all the "important" decisions is the owner. The remainder of the organization is only there to carry out orders.

This is different in the TQM organization where there is participative management and employee empowerment. In the TQM organization, it is important that everyone in the organization understand the firm's goals and objectives and the time frame for achieving them. This is because everyone is expected to be active in attaining the goals and objectives.

Lack of a formal strategic plan does not indicate a lack of direction. In fact, there may be an overabundance of direction within the firm. Everyone may have their own concept of what the firm's strategy should be. Absent a unifying vision, project teams, departments, and divisions within the company will develop and pursue their own agenda. These agendas may or may not be in harmony with the overall mission of the firm and could weaken the firm's competitive position in the marketplace.

The strategic plan must be developed by the firm's top management. When the firm formulates a strategic plan, the firm is making decisions regarding the allocation of resources. Many times, a decision to strengthen or add a particular service in the strategic process will, by default, result in other services being downscaled or markets de-emphasized. This is due to the fact that the contracting firm has limited resources.

Developing an effective strategic plan requires a thorough understanding of the firm's:

- Business practices, culture, and organizational structure
- Stated mission
- Experience and expertise
- Markets and competition
- Business and economic environment
- Social, political, and legal environment

Only the owner or upper management has this global view of the firm and its environment required to formulate an effective strategic plan for the firm. Strategic planning may be facilitated by an outside consultant or inside staff, but in the end, the planning process must be performed by the firm's top management.

PLANNING FOR QUALITY

Management's commitment to quality is a key element in the implementation of an effective TQM program within the contracting firm. Once committed to quality, the next step in the process is to determine how TQM will be implemented within the contracting firm. This step should answer a number of important questions regarding the type, extent, and timetable of the TQM program to be implemented. In addition, how corporate culture, industry structure, and other constraints will affect the implementation process must also be considered at this stage. Like the development of the firm's mission statement and strategic plan, planning the implementation of a TQM program must be completed by the upper management of the contracting firm. Implementation of a TQM program means a change in corporate culture and normally requires the dedication of significant firm resources over a long period of time. This is where upper management "walks the talk."

Deming's 14 Points and the Planning Process

Deming's 14 points are a good starting point for planning the TQM program. The 14 points identify items that must be addressed by contractors to implement a successful TQM program. The following paragraphs restate the 14 points as they apply to a contracting firm's TQM program.

Constancy of Purpose

Constancy of purpose arises from the contractor's mission statement and strategic plan. As previously discussed, the mission statement and strategic plan must address quality. The mission statement and strategic plan guide the firm in its quest for quality. As long as management adheres to the mission statement and strategic plan, there should be a constant drive for continuous improvement of the firm's business and production processes. The leadership shown by upper management will spread throughout the firm and eventually result in a positive change in corporate culture.

Adopt New Philosophy

Everyone associated with the contracting firm must adopt the new philosophy of quality and continuous improvement. This includes not only permanent employees of the firm, but also the craftsworkers working for the firm on specific projects. Total adoption of the new philosophy is necessary if TQM is to be successfully implemented in both the office and field. The adoption of the new philosophy can be facilitated by the development and dissemination of a quality policy, which gives employees and craftsworkers guidance in day-to-day matters involving quality. The development of a quality policy will be discussed later in this section.

Cease Inspection

Contractors need to stop trying to inspect quality in office and construction work. The person performing the work must be responsible for the quality of the work performed. This means that the employee and the craftsworker must be empowered and trusted to make decisions regarding quality and allowed to fix problems as they occur on their own. This is important in the field where the craftsworker's work is followed by a number of other trades. Once covered up, rework becomes time consuming and expensive. For example, imagine the incorrect size and routing of PVC raceway embedded in concrete floor slabs because office personnel failed to pass on a field directive and the foreman was afraid to stop installation for fear of holding up the general contractor's concrete placement. Quality does not come from inspection, but rather from the individuals performing the work.

Partner Suppliers

The contractor's suppliers include craftsworkers, material and installed equipment suppliers, and other contractors and suppliers. Each of these groups is extremely important to the contractor's success. As part of its TQM process, the contractor must develop and foster relationships with each of these entities to ensure continuous improvement and quality. Inputs into the contractor's business and production processes determine the quality of the work produced. Poor quality inputs normally result in poor quality outputs.

The key to quality and continuous improvement in the field is the craftsworker. To implement an effective TQM program in the field, the contractor must develop ways of bringing the craftsworker into the continuous improvement process. Without craftsworker involvement, the contractor is losing the opportunity for improved safety, quality, and productivity in the field.

Deming states that firms should end the practice of awarding business on the basis of price tag alone. At first glance, this seems to be an unrealistic expectation in the contracting business where contracts are awarded on price alone resulting in great pressure to reduce the cost of inputs to remain competitive. Additionally, many technical specifications list acceptable material and equipment suppliers and limit the contractor's leeway in this area. However, contractors should take a broader view of the procurement process and evaluate potential suppliers of material and installed equipment not only on first cost, but also on installation and maintenance costs.

By taking a broader view of the procurement process, it may be more advantageous for the contractor to select materials or installed equipment with a higher first cost. During construction, this higher first cost may be recouped through easier, more efficient installation or a more reliable delivery schedule. Following construction, the more expensive materials and equipment may result in fewer callbacks and warranty work, which will reduce overhead charges. Even though manufacturers normally reimburse contractors for the direct costs associated with the replacement or repair of defective products, contractors rarely recover indirect costs, including management and home office employee time, customer goodwill, and overhead expenses associated with correcting the problem. These indirect costs of defective products can be substantial and easily exceed the direct costs.

Wherever possible, contractors should deal only with suppliers who support their products in the field and are dedicated to quality. Factors such as technical support, ease of installation, and long-term quality of operation need to be considered. Benefits of such a policy will manifest themselves in improved customer goodwill that should result in increased opportunities for bid and negotiated work as well as reduced post-construction rework, which reduces profit.

Contractors should also work closely with other contractors and suppliers. These other contractors and suppliers include entities that provide input to the contractor's business processes as well as production processes. Examples of suppliers that support the contractor's business processes include attorneys and accountants, insurance carriers and sureties, and banks and lending institutions among others. Other than craftsworkers and material and installed equipment suppliers, tool and production equipment suppliers, engineers for design-build projects, and other specialty contractors represent some of the other contractors and suppliers that support the construction process.

Improve Constantly

Upper management must plan for constant improvement of business and production processes. Improvement of these processes is not a one-time effort but a continuous and ongoing commitment. The corporate culture must be changed so that employees and craftsworkers continually seek means and methods that will improve both the quality of the process output for the customer and the productivity of the process for the contracting firm.

Institute Job Training

There must be constant training of all individuals associated with the contracting firm. These individuals include not only permanent employees, but also craftsworkers that are associated with the firm on a project-by-project basis. The goal of these training programs is to give people the skills they need tomorrow today.

Institute Leadership

Managers and supervisors cannot just direct. They must lead! The manager must be viewed as a helper by the subordinate rather than a hindrance. Effective management means getting work accomplished through other people. Additionally, the mission of the firm must be communicated through managers and supervisors. Upper management must ensure that the primary function of managers and supervisors is to help and support subordinates in their efforts to perform their work. This means rethinking and possibly reworking the firm's and industry's traditional management philosophies. For example, the foreman in the field may need to become more of a team coordinator rather than carrying on the traditional role of "job boss."

Drive Out Fear

To have a successful TQM program and foster continuous improvement, upper management can no longer blame employees and craftsworkers for problems that arise as a result of the system. Upper management must change the focus of the organization from one that tries to affix blame for mistakes and problems to one that focuses on improving processes to eliminate future mistakes and problems.

An atmosphere must also be created within the contracting firm where subordinates feel free to communicate openly with management and supervisors regarding problems and difficulties. Subordinates must also feel free to offer suggestions for improvement of processes and be empowered to correct problems at the source. Managers and supervisors must be active and supportive listeners. They must also have both the authority and responsibility to respond positively to subordinate input and take appropriate action.

Eliminate Barriers

Barriers must be eliminated both within the support activities in the home office as well as between the home office and the field. Communication within the contracting firm must be free and unhindered by petty rivalries and turf battles. The culture within the firm must be such that everyone understands that their primary function is to support and sustain productive activity in the field. The main reason for the existence of the contracting firm is to put work in place in the field. All other business and production processes carried out within or by the contracting firm should be secondary and designed to support the field effort.

Eliminate Slogans

Slogans and directives that ask the employee or craftsworker to perform beyond his or her ability due to the system within which the work is being performed should be eliminated. "Zero punch list" or "zero defects" are goals for the system and should not be imposed on employees or craftsworkers working within the system. Slogans and directives such as these can only serve to adversely affect morale and productivity when employees and craftsworkers find that they cannot possibly achieve these objectives because of the system. Instead of impossible slogans and directives, substitute policy statements that give all employees and craftsworkers guidance as to how they should address day-to-day quality problems, improve the systems within which they work, and how they have both the authority and responsibility to control the quality of their own work.

Eliminate Quotas

The only standard that employees and craftsworkers should be judged by is the quality of their work. Quality of work, however, should not be construed as merely the goodness of the end product. Quality of work both in the office and in the field needs to include not only the fitness of the end product for use by the customer, but also the timeliness and cost effectiveness of the effort. Arbitrary deadlines, budgets, and production rates have no place in judging the quality of work.

Increase Worker Pride

Employees and craftsworkers in the contracting industry take pride in their work and want to do a quality job. Upper management must make sure that the business and production processes do not remove pride of workmanship or stifle the desire to do a quality job. In the field, the removal of unsupportive and poor craft supervision, the elimination of defective and inferior materials and installed equipment, and the removal of tools and production equipment that are not in working order or calibrated will go a long way toward increasing worker pride. Providing the employee and craftsworker what is needed to do a job when it is needed will allow the employee to focus on getting the job finished. Focusing on meeting the workers' needs will result in improved quality and increased worker pride.

Institute Total Quality Control Education

Employees and craftsworkers must receive training in TQM just as they receive continuous training on the technical aspects of their job. TQM education introduces the employee and craftsworker to the concepts and skills necessary to implement TQM in their work. This includes imparting knowledge about continuous improvement as well as interpersonal skills to improve communications throughout the organization. The employee and craftsworker must understand the importance of quality, implementation of continuous improvement, and measurement of results to the ongoing success and growth of the contracting firm.

Take Action

Management must take action if a TQM program is to be successfully implemented. This action must be in accordance with the firm's mission statement and strategic plan and include Deming's 14 points. A plan without action is of no value.

Developing a Quality Policy

Policy is a guide to managerial action. Part of management's commitment to TQM is the formulation and communication of a quality policy. The quality policy should alert employees to the fact that management is serious about quality. To be effective, the quality policy should be drafted and disseminated by top management.

A formal quality policy needs to be stated by management and complement the firm's mission statement. A formal quality policy is needed to provide consistency throughout the organization. If no formal statement of policy is made, then employees will determine by themselves what quality is or should be.

According to Philip Crosby (1979), the policy statement should be direct and to the point. For contractors, the quality policy could be as simple as, "Perform work to the exact requirements of the customer unless those requirements are formerly changed to what we and our customers really need."

For contractors, the general definition of quality developed earlier meets the requirements of the policy statement and could also be used as the quality policy within the contracting firm: "Quality is providing internal and external customers with products and services that meet or exceed their expectations and needs the first time and every time."

The quality policy could be as simple as the statement that the "customer is always right." What the quality policy cannot be is a stated goal such as "zero defects" or "zero punch list." Stated goals such as these are ineffective and can even be detrimental to the effective implementation of a TQM program. What is missing from this goal statement is consideration of the customers wants and needs as well as recognition of the fact that employees and craftsworkers must function within processes established by management. By removing consideration of the customers wants and needs from the policy statement, the message that the ends justify the means may be mistakenly sent to the organization. The focus will be strictly on the end result and not on the process where it should be. Further, the process and not the worker may be the root cause of any quality problems experienced.

Where the mission statement sets the overall goal of the organization and the strategic plan sets the course of the organization for achieving its goal, the quality policy provides employees with the guidance they need for dealing with customers on a day-to-day basis. This is even true at the construction site where the contractor can work with customers to facilitate the construction process and still comply with the requirements of the contract documents.

Establishing a Results-Based Program

To be successful, any change program must be results-based (Schaffer and Thomson 1992). This is especially true in construction where results are critical to the survival of the firm. In attempting to integrate a TQM program into the corporate culture, it is easy to become bogged down in the activities surrounding implementation such as setting up improvement teams and TQM training. An activity-centered TQM program focuses on sweeping cultural changes, large-scale training programs, and massive process innovation. Activity-based programs lose sight of the purpose of implementing a TQM program, which is to affect positive change within the organization. Activity-based programs are doomed to failure in contracting because the industry is so results oriented.

Activity does not equate to results. A results-based TQM program will bypass some of the lengthy preparatory activities and has as its goal the accomplishment of measurable results rapidly. Results-based TQM programs use external consultants and staff specialists only in limited areas where a particular expertise or knowledge needs to be taught to employees for use. To have a lasting impact and affect a change in corporate culture, the employees themselves must be involved in and be responsible for the successful implementation of TQM within the organization.

The results-based TQM program focuses on outcomes. It is important that employees see results as soon as possible and these results must be measurable. Successes, even small successes, motivate people. This is not the case where long-term, large-scale improvement objectives are used. These long-term, large-scale objectives tend to become lost in the day-to-day work. However, short-term objectives tied to current work and problems that are measurable and yield tangible results are more apt to motivate and keep the program moving. Results-driven programs focus on identifying needs and setting short-term goals to achieve those needs. Results-driven TQM programs will have a much higher probability of success in the contracting industry.

Identifying the Customer

The TQM program must have a focus and that focus must be the customer. There are two types of customers that the employees of a contracting firm must identify and satisfy: internal and external.

Internal customers are those employees within the contracting firm itself who use the work of other employees as the basis for their work. The concept of an internal customer is important in the implementation of TQM because what happens internally affects the quality of the product delivered to external customers. Shoddy and incomplete internal work affects the ability of other employees to produce a quality product delivered to external customers. For example, poor record keeping in the field on a time and expense contract affects the billing department's ability to provide the owner with an accurate and timely invoice for work completed. An inaccurate and late billing might result in the owner's making an unfounded judgment about the quality of fieldwork based on the billing department's performance.

Contractors have any number of external customers depending on the market segments they serve and the stage of the project. For example, during construction of a conventional project, a specialty contractor's primary customer is usually the general contractor or construction manager. After completion of the project, the specialty contractor's customer normally shifts to the owner for warranty and service work. The external customer may change, however, depending on the project delivery system employed by the owner on the project. In any event, the external customer's satisfaction is key to the contractor's long-term survival.

According to Juran (1988), the contracting firm and every employee and craftsworker associated with that firm has three roles: customer, processor, and supplier. Understanding the triple role concept is important because there are no pure customers and suppliers. Everyone is part of a chain taking inputs from suppliers, processing those inputs, and providing outputs to customers. The quality chain is only as strong as its weakest link. If quality is not produced by everyone along the chain, then the overall quality of the ultimate product will suffer.

Juran's triple role concept is illustrated by the simple billing example given. The billing department was a customer of the field, which supplied the raw time and material data for billing. The billing department then processed the data provided by the field and generated an invoice for the owner who is the billing department's external customer. No matter how efficient the billing department's work, its output is only as accurate as the input it receives from the field. The problem is in the process, not the billing department.

Total Quality Management Investment and Benefits

According to Crosby (1979), "quality is free." Crosby is correct; there is no additional cost for doing a job correctly the first time. In fact, there is usually a definite savings by avoiding waste, rework, and other problems associated with unsatisfactory work. There can, however, be a substantial cost associated with implementing a TQM program within a firm. If your contracting firm is currently not practicing TQM and is not quality conscious, it can take substantial training and time to integrate quality into the firm's culture. The cost of quality depends on your plans for the firm's program, how quickly you want the program up and running, who will be actively involved with the program, and how much outside training is included. You should note that the greatest cost associated with implementing TQM will be in the time spent by employees in training and other start-up activities.

The benefits of a successful TQM program can far outweigh the initial start-up costs associated with training and organizational activities. These benefits include higher employee morale, lower employee turnover, safer working conditions, greater productivity, more stable work for the craftsworker, improved competitive advantage, reduce rework, greater profit. The benefits of TQM are directly proportional to dedication of employees and craftsworkers.

Typically, it is reported that significant benefits do not accrue from implementing a TQM program for 3 to 5 years. The contracting firm, however, is smaller and more nimble than the typical manufacturing firm. Therefore, it should not take as long to see significant results for contractors. In fact, if structured as a results-based program from the beginning, concrete results can begin to be seen rather quickly. As noted previously, small results increase interest in the TQM program and momentum builds quickly within the firm.

Assembling a Quality Manual

As a way of documenting your firm's TQM program and progress, a quality manual can be assembled and used as a reference guide by employees. This quality manual can include the firm's mission statement, strategic goals, quality policy, and other important information regarding the firm's direction and policies. In addition, the manual can contain information regarding organizing and managing the TQM process and teams within the firm. Forms, checklists, and examples of the use of analysis tools and techniques can also be included. This manual should evolve over time as the firm's TQM program matures and will be especially helpful in describing your firm's TQM program to outsiders and integrating new employees into the firm.

ORGANIZING FOR QUALITY

The organizational process involves identifying the processes to be performed to satisfy the objective, assigning responsibility for those processes to particular roles, and providing the procedures to coordinate the many processes to produce the objectives. The complexity of most processes is such that individuals in a single role cannot know enough about all the parts of the process to understand where the greatest combination of changes can produce the greatest results. A team approach makes sense, involving the individuals who have the necessary knowledge about the processes to make the improvement.

The steps necessary in the process of organizing for quality vary depending on the current state of the firm. For firms with a rigid hierarchical, autocratic organizational structure and strong functional orientation, the first step is to modify the organization. Many firms have a traditional military command and control structure with rigid boundaries between functional groups. In this case, a new organization or shadow organization may need to be established.

In firms that already have adopted a team management style, the task is one of assessing the adequacy of the team, the nature of the current relationships, and the effectiveness of the communication channels to start the TQM process. Usually these firms are ready to begin the process of education, planning, and leading. With a team management approach there has already been recognition of the need for empowerment by at least a few key principals. Among modern firms that say they have a team management approach, most have team management only at the top of the organization rather than throughout.

For those firms that must establish team management, the task is more difficult. Decisions involving principals in a firm giving up their power are never made easy or without considerable thought concerning the long-term ramifications. It must be recognized that once the process has started, it will be impossible to go back. For this reason, firms may choose to operate with separate shadow organizations.

These shadow organizations must be empowered to be effective in bringing about change. These shadow organizations allow the planning for transition to proceed without completely disrupting the current operation of the firm. There is risk in this approach for both the current operation and those involved directly in the shadow organization. Where this approach makes sense is in a firm that has "old dogs" who cannot or will not learn new tricks. The current organization may be capable of sustaining the market but not capable of improving. The old organization will gradually be phased out as the new teams are brought on to create new processes and streamline the old ones.

There are cases where those with power are not so old and it must be determined whether individuals can be coached into accepting a new distribution of power. Such situations require strong leadership skills, patience, perseverance, and an effort to ensure positive outcomes for the firm. In some cases, the decision to move to a TQM approach may create a situation where some managers cannot adjust to the styles of management required and will voluntarily leave the organization or be separated from the firm. Despite the perception that may be left by Philip Crosby with the title of his popular book *Quality Is Free: The Art of Making Quality Certain* (Crosby 1979), there is a short-term price for change. Before one commits to a full-scale change, there must be the belief that survival or long-term improvement benefits outweigh this price.

Reorganization is not a prerequisite for TQM. The concept is that the organization should be a logical outgrowth of its processes. Form should follow function, not the reverse. It is not a given that change in the organization will be necessary until after the processes have been studied. These studies may indicate that to serve the customers and have a system of continuous improvement, change in the organizational structure will be necessary. There may also be cases where the study of the processes will affirm the organization's current structure and identify other areas for improvement. If process improvement is continually practiced, eventually there will be an evolution of both the processes and the organization. Form will follow function.

Organizational Structure

Quality Teams

There are several teams that make up the quality organization structure. The types of teams that are developed are depicted in Figure 8.1. These teams include the following:

- Quality management team (QMT)
- Departmental quality teams (DQTs)
- Cross-functional teams (CFTs)
- Task teams (TTs)
- Consultant team (CT)

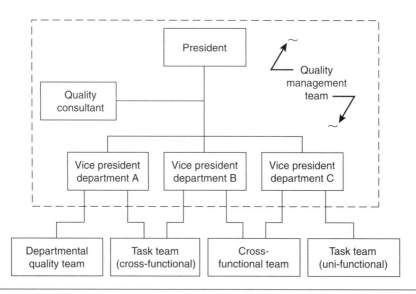

Figure 8.1 Organizing for quality

Each of these teams has different roles and responsibilities. These roles and responsibilities will be generally defined in the following sections.

Quality Management Team

The task of organizing for TQM is a top-down process. Upper management collectively will have made the commitment to quality and will in this process have created the first of several teams. This team is the quality management team (QMT). Leadership of this team is related to the leader's vision. If a shared vision exists, the leadership will be stronger since it will be bolstered by the synergy of the group toward the goal. This team may not include all upper management, but must include the ultimate power or future power in the organization if real change and distribution of power is to take place.

In small contracting firms, there may only be one or two people that represent "upper management." This does not mean that TQM cannot be successfully implemented, instead the contrary is true. The application of TQM may be easier. Since fewer individuals must commit to the change philosophy, there is greater likelihood of a closer set of common goals. Also, smaller organizations are more adaptable.

The QMT will identify the key team members to plan and guide the overall improvement process. The individuals included in the QMT should be the most powerful people in the current organization and those with the greatest knowledge of the major functions of the firm. As noted, these individuals will be giving up some of their power and must understand and embrace the necessity for change to achieve improvement. In small companies, where there isn't much upper management, this team will likely include all of the key management team.

The QMT must all commit to the quality philosophy discussed throughout this chapter. This will require training, study of additional materials, discovery and understanding of experiences with other firms, and internal discussion of the issues that directly affect the firm. The role of this team will be to provide guidance and planning for the ongoing TQM process, so it is essential that the group comes to a common understanding and focus together on the same issues.

The QMT will establish the other teams that make up the quality management organization structure. The most successful experiences with TQM have come about when this integrated approach to the organization has been successful. Some companies have assigned the responsibility for this planning to an individual responsible for quality assurance/quality control (QA/QC). This approach can work but only with the clear commitment and continual involvement and support of upper management. This individual usually does not have the power or authority to directly influence the overall organization and, as a result, the total organization may not carry the same uniform level of commitment.

The QMT will examine the broad areas of business processes and production processes that the firm performs to identify those processes that represent the "best" opportunities for improvement. These "best" opportunities may be ones with a high likelihood for success, ones that can be improved fast, or ones with significant payback. Typically, as a firm starts the TQM process, it is wise to select processes that can be successfully improved in a relative short period of time so that progress and momentum can be shown for the skeptics still remaining in the organization.

The QMT, after identifying candidate process opportunities, will select teams to address these opportunities. Depending on the process, these teams may be departmental teams, cross-functional teams, or task teams. A description of each is given later in this chapter. It is common practice for a member of the QMT to also be a part of these other teams as a liaison and to coordinate the team activities with the overall direction of the firm.

Departmental Quality Teams

Departmental quality teams (DQTs) are composed of individuals entirely from one department or major functional group. An example of this would be the warehouse department. The QMT will have identified a department or function as one that is a candidate for improvement. A team is selected from the managers and employees in the department to study a process or the processes in the department. This DQT will then take responsibility for improving the process, will have the authority and power to change the current process, and will be accountable for the results. The QMT is usually represented on this team by a member having joint membership on the QMT as well as the DQT. This joint membership provides for the expectations and commitments of the QMT to be conveyed to the DQT continuously.

This type of team is the easiest to implement and get started since it generally fits within the existing organizational structure. The DQT focuses on problems and opportunities that are solely within its internal domain or at the boundaries of the department. DQTs are a good place to start practicing process improvement principles. These represent the start of participative management. The department supervisor plays a key role in giving up some of the authority and power and also distributing the accountability for performance. Upper management commitment is key when there is a change that may involve increased investment in training, equipment, or other resources to improve total quality.

Cross-Functional Teams

The nature of many of the processes in a contracting firm is such that they span across several departments or functional groups. An example of such a process would be purchasing materials. To effectively examine these processes requires that a CFT be established to have the necessary understanding and process knowledge. These teams will be identified by the QMT to examine all of the steps in the process and make recommendations on how the process can be improved. Since these teams are examining tasks and processes across several departments, they must be empowered by upper management to bring about coordinated and integrated change in the processes. The CFT will typically have leadership from the member representing the QMT.

The formation of these teams requires upper management's commitment to the TQM process. CFTs represent the areas of greatest gain but also the greatest challenge. The team-building process is critical to these teams to get their goals reoriented from the usual departmental orientation. The risks and barriers are numerous. The need to develop open communication and trust is imperative for this type of team. Usually, results take longer to achieve. The focus of the CFT should be toward continuous improvement rather than absolute goals. It is possible that one outcome of this type of team study will be recommendations for organizational change.

Task Teams

Task teams (TTs) have long been recognized as a valuable tool for identifying opportunities for change. TT's can be organized as either a DQT or CFT. These teams are set up to examine a specific process with a specific goal in mind. The individuals assigned to this team are the process experts. There may be more than one individual who performs the same function assigned to the team. The typical practice is to capture the "best of the best" current practices and institutionalize them within the organization. Another focus may be on reducing the variability of a specific process.

When the goal is achieved, the TT is disbanded. This type of team is unique since it is given a much narrower charter and operates for a finite period of time or until a predetermined event occurs. An example of a TT would be to reduce the installation time for a particular piece of equipment by a given amount with no accidents and rework. These teams may be composed of individuals from a single department or from several departments depending on the nature of the problem defined.

This type of team has been used effectively for easily definable problems. The scope of the types of processes examined is more sharply focused and the solution boundaries are narrower. TTs can be effective for repetitive processes.

Consultant Teams

Consultants can also be part of your organizational structure for quality. The role of a CT should be one of an advisor or providing direct support for the implementation of TQM. Consultants can provide valuable training, pacing, and serve as a source of knowledge for the continuous improvement process. Outside consultants can be helpful in facilitating team building, teaching the analytical tools needed for quality improvement, and defining the firm's quality plan.

The consultant may represent a commitment to the quality process, but cannot substitute for the real commitment that must be made by the upper management team. The consultant is a facilitator to get the process going. The plan developed for implementation should include the phasing out of the facilitator as the organization makes its transition to a quality organization. The resources for sustaining a program of continuous improvement should be developed internally and viewed as a strategic resource, one that contributes to the contracting firm's competitive advantage.

Forming Teams

Each of the five types of quality teams described is designed to perform a specific function or achieve a specific quality objective within the organization. However, the types of teams formed, the number of teams in existence at any time, and the size of teams will vary drastically from company to company. In addition, the types, number, and size of teams will vary within any one company depending on the company size, time and effort dedicated to TQM, and where the company is in the implementation process.

It is not our intention to suggest that to have a successful TQM program contractors must have all of the quality teams in place and working at any one time. Other than the QMT, quality teams should be formed on an as needed basis, include only those individuals necessary to achieve the stated goals and objectives of the team, and be disbanded once the team has successfully fulfilled its mission.

MANAGING FOR QUALITY

Once the organizational structure has been decided on, the next step is to use that organizational structure as the vehicle for the implementation of the TQM program. Without an effective and dynamic organizational structure, the TQM implementation process will not be successful.

Identifying Stakeholders

Once a process has been identified for improvement, a quality improvement team is selected to begin the process of improvement. Selection of the appropriate individuals is a critical step in establishing the greatest potential for improvement and success. The key stakeholders for the process must be identified and put on the improvement team. These stakeholders include the people directly involved with performing the process; the suppliers of key materials, information, and equipment, and the downstream customer of the output who can best define the requirements.

There are several types of processes that might be examined and each will require a slightly different type of team. Many processes will involve different functions, and with these it will be necessary to have people of different backgrounds and specialties related to the overall process involved in the team.

Some processes may only involve individuals from a single specialty, function, or department, and in this case the selection of the team members is made on the basis of the experience, attitude, and individual desire for improvement. An example of this type of team would be the mechanics from the warehouse who focus on the process of changing oil and greasing machinery.

There are some processes that are studied that are specific to a single crew or ongoing work group. In these cases, the team will be composed of all of the members of the work group. An example of this would be a team assigned to improve the productivity of raceway installation on a large industrial project. This team might be made up of several experienced journeymen, two apprentices, an inspector, and a foreman who routinely install raceway.

Some processes require people from a variety of disciplines as the process crosses functional boundaries. Such a process would require a team of individuals with different types of knowledge and skills. An example of this type of team would be one that was examining the process of selecting, purchasing, and installing a security lighting system. Security consultants, lighting and electrical designers, purchasers, and installers could be involved within the boundaries of the process. In this case, it would be important to have a team with the expertise in each of these areas. This team might include an electrical designer, a purchasing agent, an electrical foreman, and journeymen electricians. Each of these is a stakeholder and part of the process.

A method for identifying the stakeholders in virtually any process is to examine several characteristics about the possible candidates. These characteristics include the control of resources (manpower, systems, equipment, and information); the level of frustration displayed (complaining, critical remarks of others, frenziedness, and preoccupation with how much time they are working); and the potential incentives to be gained by improvement. Simply stated, these are the people with the most control and the most to gain.

Establishing Roles and Responsibilities

One of the largest sources of confusion and inefficiency in a process is the lack of understanding of the participants' roles and the responsibilities assigned to each role.

When key responsibilities are not assigned, there is confusion, argument, and loss of efficiency as the accomplishment of the overall process is delayed or corrected.

Individuals must understand their role on the team to examine the process. As each is selected, they should be given a reason as to why they were chosen. This reason provides guidance as to their expected responsibilities and potential role on the team. This role may be as a leader, delegate, observer, or participant. These different roles relate to the responsibilities to provide information, lead the group, provide background, define customer requirements, identify constraints, provide linkage to other processes, represent the supplier process, control the process, or provide resources. There are many potential roles and it is vital for the group to have common perceptions of these roles as the team forms. The definition of these responsibilities may be left to the team to decide on its own in the team-building process, or it may come explicitly as part of the team chartering process as it is selected.

Fostering Employee Trust and Confidence

Upper management must develop the trust and confidence of employees to gain the full benefits of TQM. Employees must believe that they will not be hurt or put at risk of losing their job as a result of identifying ways that the process might be changed. Where the process changes are likely to result in a reduction of individuals in the process, it is important that those involved in examining the process not be the individuals at risk. Typically, by approaching the process from a quality standpoint, it is possible to avoid the problems with development of mistrust that occurs when one approaches the productivity issues from an efficiency point of view.

The promise of not losing a job is not always possible for all employees when the TQM process is started or as market conditions change. Where it is suspected that rightsizing may be necessary, care should be made to avoid such false promises of job security to those who may be at risk. Involvement in the process improvement teams should be limited to those employees who the organization wants to commit long-term employment to.

The organization should make a long-term commitment to training and "re-skilling" to keep the long-term employees continually trained as strategic employees. Individual employees must also accept responsibility for being flexible and retraining as necessary to maintain this level of job security. The marketplace dictates the changes. Seldom can a firm define the marketplace for long. The challenge for the employer is to anticipate the marketplace and adjust the human resources with training to meet the customer's demands.

Building Teams

The team members for process improvement can come from a variety of organizations. The makeup will depend on the process being addressed. Several of the processes that contractors might wish to improve will involve subcontractors and suppliers. Since contractors are the "customers" for both of these, the opportunity for cooperation is better since each wants to further its competitive advantage also. The nature of the contractual relationship may work to further cooperation or may become an issue that must be addressed in process improvement.

Key field employees should also be selected to participate on the improvement teams. There is a wide range of attitudes and relationships that exist in the workplace. When selecting individual employees to participate in a team, it is imperative that the individual have sufficient motivation and desire to see the process improved. It is also important that the individual feel secure in his or her position to have trust in the process. For hourly field employees, who are the most valuable participants in many processes, there must be a strong feeling of an organizational commitment and bond for the individual to become a full member and be active in the team. Typically, the team will come from employees who have had continuous employment for a long period of time and have an expectation or desire for continuous employment to continue.

For hourly office employees, similar principles apply as to the field employees. The individual must sense that an organizational investment has been made in them and have a desire for the process to improve. Often the individual will display a sense of frustration with the work processes and relieving this frustration will be sufficient desire in itself. The individual can also gain desire through the recognition program that goes with a TQM process and the potential for advancement and growth. Each of these potentials related to the individual should be evaluated before an individual is selected for the team.

For some processes it may be appropriate to identify other individuals to bring in for the team. These might include process specialists, such as group facilitators or consultants who would provide analytical tools for the group to employ such as time studies or statistical tools. Where there may be problems with widely variable production that appears unexplained, it may be useful to have a methods improvement specialist. An example of this might be where an electrical contracting firm is seeing a wide variation in the time required to install lighting fixtures in a high-rise building. The consultant might video tape several crews and develop aids for use in the analysis such as work flow diagrams, which will be discussed later in this chapter.

Forming and Maintaining Teams

There are several stages to creating teams for process improvement: forming, storming, norming, and performing. These stages are illustrated in Figure 8.2. Each stage must be completed before the next stage can be started. Each stage represents getting closer to the eventual goal of the team.

Tasks

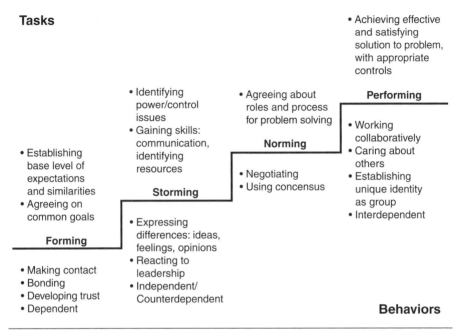

Figure 8.2 Team development stages

In the case of improvement teams, the goal is to reach a state of continuous improvement so that the team will be continually striving to achieve a higher level of performance. The stages can be achieved in a short period of time together or take long periods for progress, depending on the differences in the values and the culture of the participants, the team and communication skills of the participants, and the leadership skills contained on the team. Several key areas of training that pay great benefits as part of the overall TQM process involve team building and communication skills.

The forming process involves the team members learning more about the others in the group. Each individual brings his or her own norms of performance and feelings to the group. During this stage, there is competing behavior and testing of the leadership of the group. The forming process involves each individual becoming a team member and committing to the team. Typical behaviors during this stage of team development include:

- Making contact
- Bonding
- Developing trust
- Developing dependencies
- Attempting to define the tasks to be performed
- Attempting to define group behavior and norms

The completion of this stage is achieved when common expectations and goals are reached for the group. The feelings that the team members may have during this stage include elation at being chosen for the team, excitement, optimism, suspicion, fear, and anxiety. These different feelings lead to an emotionally charged and volatile atmosphere. This stage must be self-managed by the team itself to move to the next stage.

The storming stage involves the team members in the difficult process of understanding the tasks ahead of them and organizing to accomplish those tasks. This process is one of tremendous internal conflict as the group faces and makes decisions with uncertain outcomes based on variable experience and motivation of individual group members. The storming stage of team development includes the following behaviors:

- Expressing differences
- Reacting to leadership
- Forming and dissolving internal dependencies and allegiances
- Bickering and arguing
- Defensiveness and competition
- Jealousy
- Questioning of membership on the team

This stage is completed when the power control issues have been resolved and the communication channels have been established between members. The organization is defined and the resources that the team owns are understood by all. The feelings of the individuals during this stage include:

- Resistance to authority
- Frustration with the workload demanded
- Wide fluctuations in the belief in the goal of the group
- Frustration with the behavior of other group members
- Loneliness
- Suspicion

The norming process is one in which the team focuses on defining the real problems and developing common expectations. The norming stage involves communication of process knowledge from one member of the group to all others. The issues of criteria for determining the best alternative courses are resolved. The authority and responsibility for various decisions are with the group as a whole at the completion of this stage.

The behaviors that are exhibited during this stage are negotiating and consensus building. The group works to resolve differences by persuasion of members by other members, each recognizing and respecting the validity of others' perspectives. The group works toward achieving consensus on all major issues and satisfying members on more minor issues. The stated goals override any minor differences and the group becomes more committed to maintaining its integrity as a group than in any single particular minor issue. This stage requires mature and varied leadership styles and skills to make the transition to the next stage. There are times that it is appropriate for the leader or leaders to push, and times when the processes of information must move slowly to teach, persuade, and build consensus for an idea.

Performing is where the group starts to work in a synergistic fashion, focusing on realization of the goal. The issue of the problem-solving strategy is resolved. The roles and responsibilities are understood and each individual is executing his or her agreed functions. The group maintains its focus on the goal and is constantly monitoring its progress toward the achievement of that goal. The individual understands the responsibilities of others and appropriately becomes a responsive supplier or a customer with well-defined expectations as necessary. Each member of the group feels a strong sense of commitment to the goal and a willingness to assume additional responsibilities (even for other team members) to ensure accomplishment of the goal. The behaviors that are seen during this stage include working together, caring about others, establishing a unique identity as a group, and interdependency among members.

Empowering Employees

There are two basic, different sets of assumptions that can be made about workers. Douglas McGregor (1987) labeled these two different sets of assumptions as Theory X and Y. Theory X assumes that people are inherently lazy and must be coerced and told what to do. Theory Y assumes that people enjoy work and it is as natural as play. Theory Y assumes that individuals like having control of their own work and want to do a good job. In the last decades we have come to learn that most people fall closer to the Theory Y assumptions. Experience has shown that trust in the employee, Theory Y, will produce positive results in the work.

In terms of processes, the individual performing the process has the best idea of what could be improved and understands the relationship to upstream and downstream processes. This individual, when given the responsibility, authority, and accountability has the best potential for optimizing his or her process. The person who is in control feels the real accountability for the process. He or she feels the ownership.

Making Employees Accountable

A basic concept in organization is that responsibility, authority, and accountability should go together. Where these are not paired, there is a likely source of frustration and commitment to goals. The commitment comes from the expectation of rewards for performance, which meets the goals and the accountability for less than adequate results.

Clear communication of the goals, development of commitment toward the goals, and methods to reinforce and maintain commitment need to be addressed when goals are assigned. Techniques to communicate the problems to be addressed, methods of coordination with other processes, and constraints need to be communicated.

This whole process requires leadership. Coaching is needed to correct inappropriate behavior and rewarding is required to reinforce positive behavior. Coaching is different from the normal command and control approach that is traditional in the construction industry. Coaching involves mixing training experiences, analysis of outcomes, exploration and discussion of causes and effects, and transfer of experiential knowledge from one individual to another.

The coaching effort can be viewed by the employee as an organizational investment. The time put in by the coach represents an investment made in future performance and knowledge imbedded in the individual. Coaching creates a bond between the employee and the coach based on mutual respect and commitment to the employee's career. This concept is not a foreign one to contractors, since this is the essence of the journeyman-apprentice approach to the technical training. This concept can also be continually applied to management-employee relations to create a long-term seamless relationship with key employees.

Recognizing Contributing Employees

A basic precept of TQM as a management philosophy is that "We all desire recognition for the things we do well." We want that recognition in different forms. Some individuals want to be recognized in front of their peers. Others want to receive recognition in another setting, such as in the family or other social group.

Examples of recognition programs include gifts, money, and other outright incentives. Recognition can also be in the form of a verbal statement in private or in public for an employee's contribution to the team effort. It is particularly useful to reward and recognize at the team level, rather than at the individual level, since this further promotes the team concept.

The risk in any reward or recognition system is the problem of jealousy, loss of team orientation, and negative competitive behavior. These problems are normally the result of a poorly thought-out incentive program. Care should be exercised when developing such programs. In particular, the rewards and recognition should be meaningful to the employee and team, the system should not do harm to the motivation of other teams, and the recognition should be for performance that truly advances the organization toward realization of its stated mission.

IDENTIFYING IMPROVEMENT OPPORTUNITIES

In order for the continuous improvement process to get started in the contracting firm, improvement opportunities must be identified for process improvement teams to work on.

The quality improvement team provides leadership and guidance for the selection of improvement opportunities. These opportunities can come from many parts of the contractor's organization. The purpose of TQM is to improve the contractor's processes, which in turn results in the end product or service being improved.

Opportunities can be classified as either business or production processes. Both are required in any firm and depending on the boundaries established for a candidate process, some may contain both business and production processes.

A contractor has many of the same business processes as other businesses. The detailed steps and relationships between the process steps are quite varied. A business process is one that supports the production process and supports the overall objectives of the organization.

There are hundreds of business processes that are a part of being a contractor. These business processes consume a large percentage of the resources that go into the final product. The efficiency and effectiveness of the business processes can have a significant impact on the resources consumed in the production processes. At many points in the overall scheme of work flow, the business processes interact with the production processes. Problems that exist in the business processes may actually surface as symptoms in the production processes. Typically, our feedback sensors are set up to detect problems that occur in the production process or downstream with the customer.

Each firm can develop a list of its own business processes. Table 8.1 shows a few of the typical business processes that any contractor might have.

Contractors have a number of production processes that represent the delivery of the key services and products that the customer desires. The output of a production process is the service or product that is ready for delivery to the external customer.

For an electrical contractor, for example, production processes include the installation of conduit, installation of equipment and fixtures, termination of conductors, checking of circuits, and many more depending on the contractor's specialty. These processes are the activities that lead to the product that the contractor is paid to perform. Assuming that the specifications are achieved, all contractors end up with the same product. The thing that differentiates contractors from each other is the way, or process, that they use to achieve their product. It has been shown that tremendous amounts of a craftsworker's time are spent waiting, handling materials, and other nonproductive activities. The reasons are numerous, but the solutions all lie in examining the processes.

Table 8.1 Typical business processes

Area	Processes
Estimating and sales	Marketing Cost estimating
Personnel	Personnel search Orientation and employment sign-up Recording/Governmental reporting
Finance and accounting	Ledger control Cash management Payroll Financial planning
Purchasing	Supplier selection Expediting Inspection
Operations management	Planning and scheduling Timekeeping Production control Job start-up

Contractors have many opportunities to improve the firm's "bottom line" through improvement of production processes. These processes include many interface points with business processes. Contractors must recognize these interfaces when looking for improvement opportunities. These interfaces can be the "bottleneck" that limits improvement. The key to successful improvement projects lies in careful definition of the problem or opportunity.

Screening Opportunities

There may be many improvement opportunities that can be selected by the quality improvement team. As a team, this group should have a full perspective of all of the functions and processes that are at work in the firm. Since these individuals are responsible for different functional areas, this team must develop a full and common understanding of all of the processes before selecting processes for improvement teams to work on.

The improvement process will require resources and effort. It is not free. The quality improvement team is accountable as a group for these resources, and must identify opportunities that will provide an appropriate rate of return in the long run. As the opportunities are identified, this team must fully understand that the goal is not the improvement process, but the results that are derived from the improvement process. The team must have full and complete knowledge to make the process selection decision. The costs and benefits must be carefully weighed in the selection process.

The first step in the quality improvement process is for the quality improvement team to completely understand the functions and processes. In a large company, the team members are usually knowledgeable in their own areas of responsibility, but may not be knowledgeable in other areas. In some cases, the organization may have so many levels of management that an individual may not fully understand the detailed processes behind the functions for which he or she is responsible. In small companies, there is usually some division of responsibilities, at least between the business and production processes. In these companies it is important that the team understand both types of processes as well as the interfaces between them.

An effective starting point is to begin with the current organization and individually develop a list of all of the functions and processes that are performed within each unit. This will get the team to begin using the same language and start to understand that many of the processes are cross-functional and extend beyond organization boundaries. Starting at the summary level, an effort should be made to develop simple diagrams for the various functions. Then each team member should begin the process of flowcharting several of his or her major processes and sharing these with other members of the quality improvement team.

The proper way to examine improvement opportunities is like any other investment opportunity. Each firm has a limited set of resources that can be invested. These resources, particularly the human resources, have a finite amount of knowledge, creativity, and energy to use for both production and for this investment in the future. The quality improvement team must decide on the appropriate level of investment and, therefore, the pace and expectations for improvement.

The quality improvement team should examine both the estimated investment of resources and the potential benefits. The resources are usually a combination of training costs (team building, quality improvement basics, and some technical training); improvement team processing costs (the cost of meetings, data gathering, and internal and external inputs); and the costs of implementation. Although these are uncertain, an attempt should be made to estimate and monitor these costs.

A good analogy is the taxi. Even if you go a short distance, there is a basic cost. As you go farther, the unit cost goes down. Going still further, the meter keeps running and you must feel confident that you are getting closer to your destination or you'll want to stop and catch another taxi. Experience has shown that TQM has many parallels with a taxi trip. The team will get better at estimating the costs as they develop experience with several processes. Thus, it is better to start in a limited way and add improvement projects as the firm gets experience with the approach.

The quality improvement team also must weigh the potential returns that might be realized with successful implementation. The improvement team will typically examine the mission of the firm for direction; examine the areas where the symptoms of problems are appearing (complaints, lack of competitiveness, or others), and the areas where significant resources are currently directed. The team must collect and share all the performance data that are available. In some cases, the team must develop data on the performance of processes. This typically involves developing customer questionnaires, customer interviews, and focus groups. The team should look at both internal and external customers in this process. The group should also carefully examine the nature of the problem to determine that the apparent problem is not just a symptom of a deeper problem in a different process.

Process improvement is like any other project; it will consume resources and requires control of the cost and time for it to achieve a reasonable return on investment. A plan is necessary. The quality improvement team has an expectation as to when the return will be realized and this needs to be transmitted to the process improvement team. The quality improvement team will also need to identify the order of processes to be examined when several are going to be studied. Since there may be relationships between processes, these ties should be recognized. Some process improvement opportunities will identify others. This situation should be recognized and utilized. Since resources are limited, the quality improvement team must pace the improvement projects recognizing resource constraints.

Analysis Tools and Techniques

There are a variety of basic tools that can be used to portray data and help team members understand their meaning. These tools are merely communication aids for teams that are analyzing processes. The teams must apply creativity and leadership to identify the actions necessary for continuous improvement.

Cause-and-Effect Diagrams

Cause-and-effect diagrams are also known as fishbone or Ishikawa diagrams. These diagrams are tools that help identify the root causes of problems that are being analyzed. When complete, these diagrams tend to resemble a set of fishbones. These diagrams can be used to identify a process that is a problem or a critical activity or task within a process that is a problem. An example of a cause-and-effect diagram that might be used to analyze cable-pulling productivity on an industrial project is shown in Figure 8.3. The process of developing cause-and-effect diagrams involves the following steps:

- Identify a potential problem to be studied.
- Identify potential causes, typically grouped into materials, methods, manpower, and machinery. Diagram these as major branches.
- Continue to trace the causes of the causes on these branches.

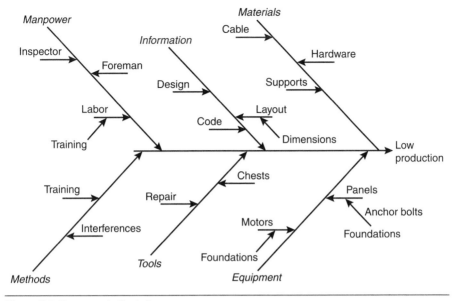

Figure 8.3 Cause-and-effect diagram

Flowcharts

Flowcharts are useful tools for illustrating the various steps of a process. Flowcharts can distinguish between the various types of steps in a process, the relationships between these steps, and alternative decision paths within the process. The purpose of flowcharts, as the name implies, is to show process flow. This flow may be information, a product, or a resource. Flowcharts illustrate the direction of movement and show the major events in the process.

There is a set of standard symbols used in most flow-charting. These symbols are shown in Figure 8.4 along with their application. The symbols can be put together to illustrate a variety of processes. Once the process is shown pictorially, the improvement team can begin to identify problem areas, examine alternative flows and subprocesses, examine the importance of steps included, or consider adding additional steps. The flowchart provides a common process language for analysis. An example flowchart illustrating the receiving process for fixtures on a large commercial project is shown in Figure 8.5.

Histograms

Histograms show the frequency of occurrence of each process attribute being studied. The horizontal or x axis of a histogram lists the attributes of the process being studied. The vertical or y axis of the histogram shows the frequency of

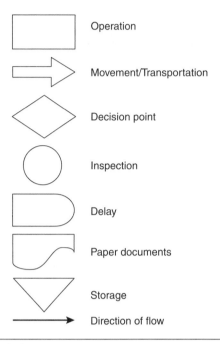

Figure 8.4 Flowchart symbols

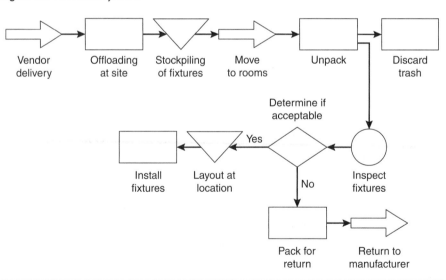

Figure 8.5 Flowchart

occurrence for each attribute listed along the horizontal axis. In Figure 8.6, a histogram is used to illustrate the frequency of various markups over the period of a year for a contracting firm.

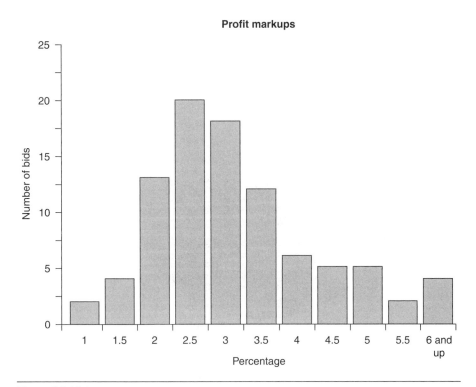

Figure 8.6 Histogram

Pareto Charts

Pareto charts are used to compare percent values for several process attributes being studied. The purpose is to identify those attributes that have the greatest potential impact on the process and, as a result, represent the greatest opportunity for process improvement. Pareto's Principle is the basis for the chart and is sometimes referred to as the 80-20 rule. Simply stated, the 80-20 rule says that 80 percent of the problems associated with a process result from 20 percent of the process steps. By analyzing a process using a Pareto chart, the improvement team can focus on those process attributes that cause most of the problems. This improves the team's efficiency and avoids expending resources on improvements that will produce little benefit. An example of a Pareto chart used to study the time spent by a termination crew is shown in Figure 8.7.

Scatter Diagrams

The scatter diagram provides a picture of the relationship between two variables. The name comes from the appearance of the picture. The stronger the relation-

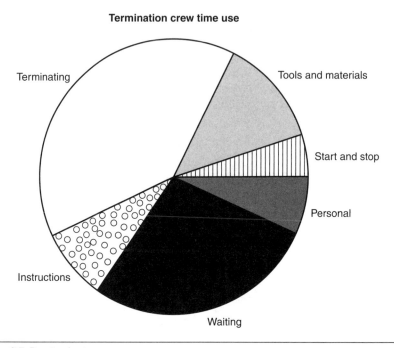

Figure 8.7 Pareto chart

ship between the two variables, the less scatter there is to the data points. The scatter diagram then can be used to develop a quantifiable relationship between the two variables. The scatter diagram can be used to show the relationship between several variables of a construction process as illustrated in Figure 8.8. This figure shows the relationship between temperature and performance for cable pulling, termination, and instrument installation.

Run Charts

Run or trend charts provide a timeline for comparing data about a variable of interest. The run chart typically tracks a trend or the change in a variable. Typically, a run chart is used to identify the seasonal variations in some business activity. Another application is to chart the variation over the days of the week or month for a particular activity.

An example run chart is shown in Figure 8.9. This chart shows the daily variation in productive time during the course of a workweek. From this chart it can be seen that direct work is highest on Tuesday and lowest on Monday. After Tuesday's peak, daily productivity steadily decreases through Friday.

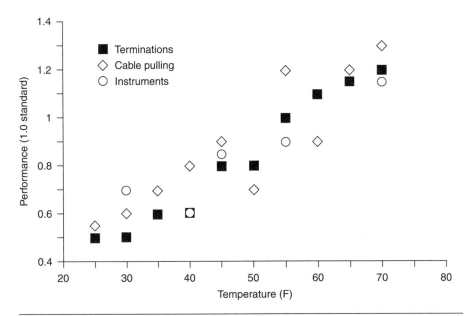

Figure 8.8 Scatter diagram

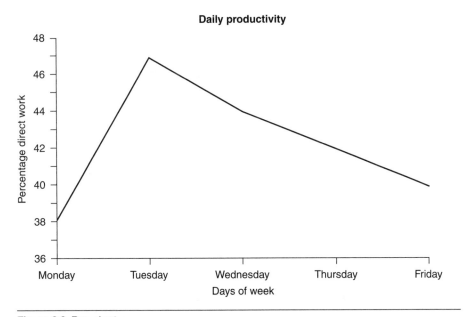

Figure 8.9 Run chart

Control Charts

Control charts are simply run charts with an upper limit (UL), an average (AVG), and lower limit (LL) shown on the chart. These reference lines provide the basis for control. These charts are often referred to as statistical process control (SPC) charts.

The reference lines are determined by letting the process operate without interference and collecting data. The data are then used to determine the UL, AVG, and LL of the process using statistical methods. Control charts assume that variation can be expected in any process. The purpose of the control chart is to detect when major variations occur in a process that cannot be explained by normal random variation. Persistent or periodic variations outside the control limits indicate that the system is out of control and corrective action needs to be taken.

An example process control chart for studying instrumentation installation productivity is shown in Figure 8.10.

Benchmarking

Benchmarking is nothing more than making a comparison between two firms operating in the same market. The challenge in benchmarking is to select the

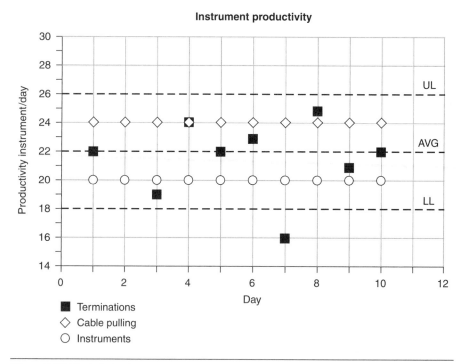

Figure 8.10 Control chart

appropriate benchmarks for comparison. The guiding principle for benchmarking performance is to select a representative firm for comparison that would perform the work if you did not. As an example, if you were an excavation contractor you would benchmark your performance against another excavation contractor, not an electrical contractor. In this way, a true comparison of the best of the best is made.

Benchmarks give a good idea of the potential areas for establishing competitive advantage as well as an objective measure of expected performance. Benchmarks establish a visible target or destination for the quality journey. Benchmarking must be continually practiced since the competition is dynamic.

Establishment of the benchmark measures typically comes from the strategic planning process where the strengths of a firm are identified. If a firm believes it is the best at something, such as installing and maintaining life safety/security systems, a benchmarking study is an appropriate way to verify this and track relative quality over time.

Often a firm finds that the things that it's best at may not be that important to the customer. Benchmarks should be established that focus on the customer. This is a useful outcome of the exploration phase that a team goes through to identify improvement steps in a process.

Other Tools and Techniques

There are other tools that are adaptations of the basic tools discussed. These include such things as crew balance charts, decision mapping, and construction process simulation. These tools and techniques all represent ways that data can be turned into information for the purpose of analyzing and improving a process. These advanced tools and techniques will not be covered in this chapter.

IMPLEMENTING TOTAL QUALITY MANAGEMENT IN THE FIELD

The greatest opportunity for continuous improvement for contractors lies in field operations. Field operations represent the contractor's production process. Putting work in place in an efficient manner is what contractors are all about. If contractors are not effective at putting work in place, then they will not be successful in the marketplace.

The construction contract establishes the requirements that define quality for the contractor. In particular, the project's plans and specifications establish the technical requirements that must be met. Under this definition, the sole criterion for judging the quality of work in place is compliance with the contract documents.

The concept of quality as it applies to construction is not always understood. Quality in construction should not imply that the contractor employing TQM is to be held to a higher standard of care than that specified in the contract documents. Quality in the field is not a degree of goodness. In a competitive bid situation, contractors must base their bids on the requirements set forth in the project plans and specifications. Any enhancement to these requirements will affect the contractor's ability to compete in the marketplace and may not provide the owner with what is really wanted or needed.

There are two dimensions of quality that must be addressed in the field:

1. Continuous improvement of construction processes that aim at eliminating wasted effort, materials, and equipment usage
2. Elimination of nonconforming work which has as its objective zero defects or punch list

These two dimensions are not mutually exclusive and both contribute to the success of the project.

TQM is about the effective and efficient management of processes. Processes in the field are the construction means and methods that contractors employ to complete the work. Every job is a process that has both inputs and outputs. The contractor's means and methods represent a process where inputs such as labor, materials and installed equipment, and production equipment are transformed into work in place. Figure 8.11 illustrates the construction process.

Construction means and methods can also be thought of as a production function that relates inputs to outputs. The production function specifies the amount of output that can be produced for a given combination of inputs. The production function is determined by a number of factors that include the production factors employed by the firm, level of technology, and management skill. In construction, the output of the production function is work in place. The input to the production function is the factors of production: labor, materials

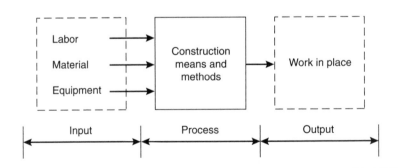

Figure 8.11 Deterministic production function

and installed equipment, and production equipment. The goal is to maximize the work in place for a given amount of inputs or, alternatively, minimize the quantity of inputs necessary to achieve a given amount of work in place.

The production function represented in Figure 8.11 is deterministic. By deterministic we mean that the contractor will always get the same output for the same inputs using the same means and methods. In other words, the process is reproducible. The deterministic production function is what you might expect for a manufacturing process.

Both manufacturing and construction assemble elements to create a finished product. However, the construction industry has many variables that are not present in manufacturing and must be dealt with. For one thing, the project location is never the same. For another, personnel are constantly changing. Different materials are used on different projects. Weather is also a significant factor that affects productivity. Figure 8.12 expands on Figure 8.11 and illustrates a production function that takes these variables into account. The output of this model may or may not be the same for the same mix of inputs since the process is at the mercy of many outside variables. Outside variables affecting the production function in Figure 8.12 include physical factors, climate and weather factors, social factors, economic factors, and legal factors. In addition, the quality, completeness, and requirements of the contract documents also affect the contractor's production function.

These outside factors that affect the contractor's production function are why traditional TQM methods employed in the controlled environment of manufacturing are not always applicable to the construction process. Contractors must try to control these outside factors through effective planning.

Despite the differences between construction and manufacturing, the principles of TQM are applicable to construction because it is a process. TQM can be successfully implemented to significantly improve productivity and safety, which will ultimately reduce costs. Any TQM initiative in the field, however, must be tailored to the transitory nature of the workforce and the non-repetitive nature of the work due to the uniqueness of projects.

Factors for a Successful Implementation in the Field

Work management in construction is the vehicle for the successful implementation of TQM in field operations. Work management is the effective use of available resources within the project environment to achieve predefined objectives of quality, cost, and schedule. Through effective work management, contractors can achieve results in the field despite the obstacles of a transitory workforce and non-repetitive work. Effective work management requires that the contractor analyze all facets of the job to determine how the job can be completed with less effort, with a lower cost, with greater safety, and with more speed.

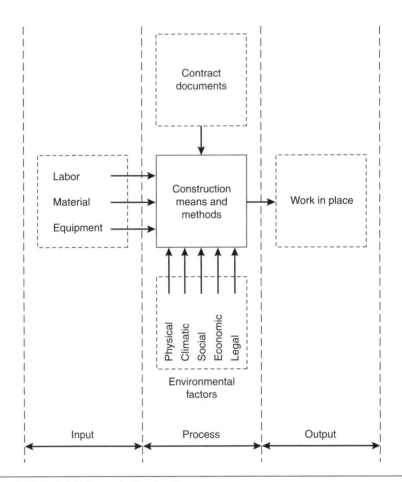

Figure 8.12 Probabilistic production function

Peter Drucker (1974) states that productivity is the first test of management's competence. Getting the necessary resources and putting those resources to work is only the beginning. The ultimate task facing contractors is to make those resources as productive as possible. The continuous improvement of productivity is one of the contractor's most important jobs.

The model for work management in the field should be the Shewhart (1986), Deming (1986), or PDCA (Construction Industry Institute 1990) cycle. PDCA stands for plan, do, check, and act. The PDCA cycle is a systematic procedure that focuses on correcting and preventing defects to improve methods and procedures incrementally. The objective of the PDCA cycle is to remove the root causes of problems and continuously improve methods and procedures. The intended results of the PDCA cycle are to (1) achieve continual improvements in methods and procedures and (2) ensure that improvements already made are maintained. A graphic representation of the PDCA cycle is shown in Figure 8.13. Continuous improvement is achieved through the PDCA cycle as follows:

- PLAN: Design improvements for present practices
- DO: Implement the plan on a small scale
- CHECK: Verify the results of the plan
- ACT: Corrective action on the process, standardize the process, and feed forward to the next plan

The PDCA cycle was originally developed for manufacturing where there is the opportunity to develop and test prototype production processes before full implementation. This is not always possible in construction where the demands of a project often require quick and decisive action. A condensed PDCA cycle where the Do and Check activities are combined into one Review activity can produce a three-phase process improvement model. It is seldom possible to implement an improvement plan on a small scale and then verify the results during construction. However, the plan can and should be reviewed in detail once formulated by those who must implement it to ensure that it is sound.

The continuous improvement process involves maintenance, improvement, and innovation as shown in Figure 8.14. Continuous maintenance and improvement are important to maintaining and improving competitive advantage. However, innovation is critical to substantial gains in competitive advantage.

Field Implementation Model

Construction is a process where the factors of production are transformed into work in place through construction means and methods. It is the effective management of these means and methods that will result in improved productivity, increased safety, and reduced rework.

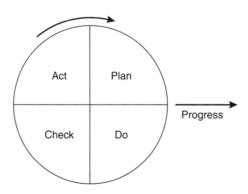

Figure 8.13 PDCA cycle

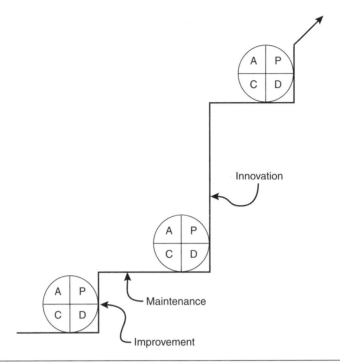

Figure 8.14 Continuous improvement cycle

Casten and Howell's model of the construction process requires that the contractor supply a number of elements to the crews in the field in order for those crews to be productive (Olson 1992). These elements, reordered and revised for the construction process, include the following:

- Information
- Skilled craftsmen
- Materials and installed equipment
- Tools and production equipment
- Time to perform the work
- Place to perform the work

A graphic model to illustrate the construction process is provided in Figure 8.15. As can be seen from this model, everything flows from the contractor through the foreman and crew ending up as work in place at the site. Contractors are responsible for developing and improving the system of putting work in place at the site. The craftsman is responsible for performing work at the site within the system provided by the contractor. The foreman and crew cannot be held responsible for poor quality that results from a dysfunctional system.

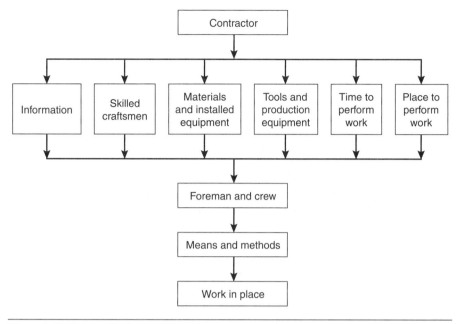

Figure 8.15 Construction process

Information

The foreman and crew in the field need information to complete the required work. This information is usually included in the project plans and specifications. However, there is not always sufficient detail in the plans and specifications to allow the foreman and crew to proceed with the work unimpeded. Lack of detail, errors and omissions, questions of coordination, absence of material and equipment installation details, and other factors can affect the foreman and crew's ability to complete the work in a timely and cost-effective manner. Designing and detailing in the field by the foreman and crew can be expensive and adversely affect productivity.

Contractors need to evaluate the adequacy of the design documents for use in the field. Where clarification and additional information is needed for the foreman and crew to proceed efficiently, the contractor needs to obtain that information from the owner, general contractor, other subcontractors, or material and equipment suppliers before the lack of information affects productivity. Where information cannot be obtained from outside entities, the needed information should be developed in-house and provided to the field to avoid slowing construction activity.

The additional or supplemental information required may be nothing more than translating the requirements set forth in the contract documents into language that is understood by the foreman and crew. For example, a specification requirement to torque lugs to the manufacturer's specification may require the contractor to obtain the necessary information from the equipment manufacturer and provide it to the foreman and crew in the field. Needed information such as this provided in advance of the crew's need will reduce idle time, prevent rework, and improve morale.

Skilled Craftsmen

Craftsmen with the skills needed must be provided if the foreman and crew are to be effective. Additionally, sufficient numbers of craftsmen must be provided to complete the work efficiently and on schedule. For example, in electrical construction, one craftsman can install lighting branch circuits by himself. However, a lot of time will be wasted pulling and terminating branch circuit conductors if the craftsman has to work both ends without help.

Materials and Installed Equipment

Materials and installed equipment must be provided to the foreman and crew when and where needed. On poorly planned projects, a great deal of production time is lost waiting for material and installed equipment as well as multiple handling. In addition to the proper quality, the quantity of materials and installed equipment also needs to be addressed. It may be in the contractor's best interest to provide better materials and installed equipment from those specified. The labor savings associated with the installation of the materials and installed equipment may well offset the additional cost of the better product. In addition, off-site prefabrication of assemblies by craftsmen in a controlled environment such as a nearby warehouse may result in higher quality and greater safety for the craftsman during times of inclement weather.

Tools and Production Equipment

Providing needed tools and production equipment to the foreman and crew when and where they are needed is as important as the supply of materials and installed equipment. The tools and production equipment should also be in working order and calibrated to improve safety, productivity, and morale. For example, if a lug has a specified torque value associated with it, then a torque wrench needs to be provided. If two scissor lifts are required to improve crew productivity installing high-bay lighting fixtures, then two scissor lifts should be provided.

Time to Perform the Work

Sufficient time must be provided for the foreman and crew to perform the necessary work. Lack of sufficient time to perform the work can result in a degradation of quality and lower morale at the site. Also, if extended overtime or multiple shifts are required, lower productivity due to fatigue and other factors needs to be factored into management's expectations and understood.

Place to Perform the Work

Adequate space must be provided for the foreman and crew to perform the work safely and efficiently. Not only does there need to be adequate physical space, but the space also needs to have sufficient light, be well ventilated, conditioned to a reasonable temperature whenever possible, and clean.

Organizing for Quality in the Field

An organization chart for organizing for quality in the field is shown in Figure 8.16. As can be seen from this figure, upper management directs the field quality effort through two types of teams on each project: project team and field teams.

The organization for successfully implementing TQM in the field is similar to the way most contractors organize their construction effort now. There are no new personnel or titles added. In some firms, a single individual may be responsible for several or even all of the support roles associated with the project team. This lean organization is important for the following reasons:

- Construction is competitive and only the largest contractors could afford to have full- or part-time quality personnel on staff and adding to the company's overhead.

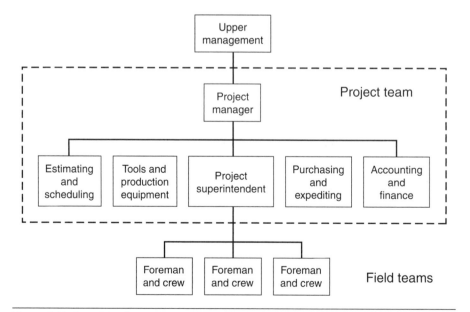

Figure 8.16 Organizing for quality in the field

- If the implementation of TQM in the field is going to be successful, it has to be executed by those actually responsible for and doing the work. Quality must be an integral part of the firm's culture.
- The contractor's field TQM effort must be organized around each specific project undertaken. Each project is unique and offers different problems and challenges that need to be overcome.

Project-Specific Teams

Each project should have a project team designated that is not only responsible for the project in the traditional sense but also quality. The size of the team depends on the size and complexity of the project and could consist of just one individual. Upper management should formally state the mission of the project team. An effective mission statement might be as simple as "Complete the project on time, within budget, and in accordance with the contract documents."

Upper management can enhance this mission statement with other project-specific goals and objectives. However, while specific about the outcomes wanted, the project mission statement should avoid specifying the means for attaining the stated goals. The means for achieving the project mission should be left to the project team. The mission statement should be legitimate, measurable, and attainable.

The project team should be composed of managers who control the resources and services needed to complete the project. The project manager should be the team leader as shown in Figure 8.16. As noted earlier, one manager may be responsible for two or more support functions. The number on the team is not important. The key is that whoever the support team members are, they realize that they all are part of the project team and responsible for supporting the foreman and crew in the field.

Field Teams

Total teamwork is necessary for the successful implementation of TQM in the field. The key to the successful implementation of TQM in the field is the active involvement of the craftsmen actually performing the work. This can be accomplished by setting up field teams as shown in Figure 8.16. Contractors can use journeymen meetings to identify and correct field problems affecting the project. These journeymen teams can study and suggest solutions to field problems involving the elements previously discussed. Loosely structured, these team meetings encourage participation.

Before contractors can begin to improve construction processes, they must first know where the problems are that are affecting the craftsmen's ability to produce quality construction. This requires that the contractor set up mechanisms for obtaining meaningful input from field personnel and crews. This is where the craftsman, even if it is a transitory relationship, can provide invaluable input. By asking and observing, the contractor may be able to identify many of the problems experienced by the craftsman with the system. Once these problems have been identified, the field and project teams can work together to solve the problem within the constraints of the project.

Craftsman Involvement

Quality in the field must start with the craftsman. The craftsman is both trained and experienced in his or her work. Quality must be managed through the skills of the craftsman. The responsibility for the quality of work in place needs to be delegated to the craftsman. If the craftsman becomes responsible for the quality of the work in place, then management must provide the craftsman with the resources needed to complete the work as well as to empower the craftsman to take corrective action when needed to ensure quality. Under these conditions, corrective action will take place as the work is in progress, reducing rework.

Contractors cannot hold the craftsmen accountable for quality if the craftsmen are not empowered to control the work in progress. The benefits of empowering craftsmen in the field include a shorter feedback loop and earlier response to quality problems, as well as a greater sense of participation and ownership of the work by the craftsmen.

Establish a Dialogue about Quality

Feedback from craftsmen is extremely important. Typically, the craftsman has direct contact with a number of outside entities at the project site on a daily basis. These outside entities include personnel working for the project owner or other contractors. As a result, the craftsman knows the process and what is needed in terms of information, tools, and materials. To the extent possible, the craftsman should be involved in the selection and development of means and methods for accomplishing the work. Craftsmen need to be able to control their own work.

Self-Managing Teams

There has to be a change in traditional management attitudes and practices in the field to successfully implement TQM. As previously noted, the contractor is responsible for developing and improving the system and the craftsman is responsible for performing work within the system. For TQM to be successful in the field, participative management must be implemented. The system is never perfect and can always be improved. The best people to consult as to what is wrong with the system and how the system can be improved are the craftsmen.

The foreman needs to become a team coordinator. Foremen need training in participative management, group facilitation, and problem solving. In addition, the foreman needs to understand planning and scheduling and its relationship to the achievement of the project goals and objectives.

Planning for Quality in the Field

Planning is the starting point of the PDCA cycle. Planning is the key to continuous improvement. Lacking any one of the elements needed for effective production will keep construction from proceeding. More importantly, however, insufficient quantity or poor quality of one or more of these elements will also adversely affect the productivity of the foreman and crew in the field.

It is not uncommon to start work without a detailed plan. The justification is that the activity has been performed before on numerous other projects and, therefore, can be performed again in exactly the same way. Rather than formal participative planning, the contractor forges ahead performing the activity exactly as it has been performed in the past. The contractor proceeds in this way because it is the way that it has always been accomplished in the past. What is missing is the reason why so many people in the construction industry say that management techniques developed for manufacturing won't work. Every project is different. Therefore, if this is so, then planning is critical to the contractor's success.

Without reexamining how work is to be performed on a project-by-project basis, continuous improvement is not possible. A reexamination of how even the simplest, most repetitive activity is to be performed on a particular project should be completed if there is to be improvement or innovation in the field. In fact, just maintaining the status quo requires periodic checking and examination of means and methods. If you are not moving forward, then you are probably moving backward.

If contractors were to examine their work, they will probably discover that most of the work lies in a small number of activities. Improving productivity on these repetitive activities probably has the greatest potential for overall improvement, yet these activities receive the least attention since they are so routine.

The need for planning is often confused with the criticality of the activity. There is no doubt that a new activity not performed before needs to be planned and executed with great precision. However, when we are looking at improving overall field operations, we also need to focus on the everyday activities that make up the routine daily work at the site. These everyday activities are the activities where the greatest opportunity for improvement lies for contractors.

Planning and continuous improvement do not only occur through technological innovation. Studies of commercial and industrial construction have shown that a significant portion of a craftsman's time is wasted due to not having the necessary quantity or quality of information, material and installed equipment, or tools and production equipment needed to perform the work. Contractors need to ensure that the craftsman has what is needed when it is needed through the planning process.

Planning must be performed on a variety of levels by the contractor. These levels include project-level planning, intermediate-level planning, and activity-level planning.

The project plan needs to be concerned not only with time, but also with resources and costs. In putting together the plan, the contractor needs to involve not only the project team but also subcontractors and suppliers. Also, the superintendent and foremen who are going to perform the work should be involved whenever possible. Developing the project plan can be a team-building exercise. Everyone should feel like they contributed and were part of the planning process.

Once developed, the project plan should be implemented. At regular intervals or when unexpected events occur, the project plan should be reviewed and modified to reflect current and anticipated conditions.

The project plan provides the framework for developing intermediate plans in the field. The implementation of the project plan requires that individual work activities be started, performed, and completed to ensure that the project is completed on time and within budget. To ensure that the project objectives are achieved, intermediate-level planning must take place at the site. The purpose of intermediate-level planning is to identify the activities that must be accomplished in the following days, weeks, or months to complete the overall project on schedule and within budget. Intermediate-level planning is sometimes referred to as short-interval scheduling, which includes both look-ahead and production schedules.

Activity-level planning is performed at the crew level and is sometimes referred to as construction preplanning. The goal of activity-level planning is to ensure that the goals set forth in the intermediate-level plan are achieved. Activity-level planning is also considered micro-planning because this planning is carried out at the crew level where the work is actually accomplished.

The goal of activity-level planning is to get the craftsman the appropriate quantity and quality of what is needed at the correct place and time. This includes not only material and installed equipment and tools and production equipment, but also information as to what has to be completed and how it should be completed.

Activity-level planning results in a detailed plan as to how a particular task will be accomplished. Activity-level plans involve the crew and replace the foreman's informal daily plans. Activity planning requires that the activity be thought through by the crew to determine the who, what, where, when, and how.

Activity planning represents a detailed plan as to how a particular task will be accomplished. Activity plans involve the crew and require thinking through the task before undertaking it. Activity plans lead to better coordination and identify and correct problems before they occur.

MEASURING RESULTS

The business environment today requires accountability and discipline. Today's decisions must be backed up by hard data and facts. The decisions to change business and production processes deserve to be supported by objective data and the results should be proven out with hard data.

To manage a process, one must be able to measure the steps in the process. The focus of any effort to change a process should and must be on results. The only way to develop and maintain this focus is by developing objective measures of the results.

Management by Fact

It is easy to fool oneself into believing that there is control of a process or even improvement in a process if the only input is perception based on a biased attitude and level of effort toward a goal. Many of the actions that we take in our businesses are based on faith and the results are simply perceptions. In fact, there is a marketing philosophy that suggests that perception is reality in our customer's eyes. This may be true, but our customers are increasingly looking to quantitative measures of results to base their perceptions on. It is a short-term view that suggests that we can effectively control our processes with perceptions, which are not based on facts.

The principle of management by fact or objective results is an important one for convincing our customers and for developing a solid foundation for changing our processes. We must develop facts that are the appropriate measures of our performance at various points in our processes, and at points of exchanges between suppliers and both internal and external customers. These various facts together create a perception with a consistent discipline to compare the performance over time.

There are several strong reasons for developing measurement or feedback mechanisms. The basic need comes from the need to understand what is occurring in our company or on our job. The feedback mechanisms provide information for evaluating the need for change to occur and evaluating the impact of change.

The feedback also provides a way to track process improvements and ensure that gains made are not lost. As a process is reengineered to provide for changes, it is important to get a sense that the anticipated effects are coming about as a result of the changes.

Feedback mechanisms are useful for identifying and correcting out-of-control conditions. The data can show where there are wide fluctuations in the performance of a process. Feedback over time can clearly show this situation and wide fluctuations are an indication of an out-of-control process.

The feedback mechanisms recognize the important parameters and provide a way of setting priorities for improvement. The feedback can be used to identify when additional training is needed. The feedback mechanisms provide a tool for planning to meet new customer expectations.

The feedback mechanisms are vital to developing individual commitment to a process. The feedback provides the feeling of accomplishment. Keeping score by some means makes the process more like play, a game. Individuals are more responsive and adaptable if measures of performance are available to them. This approach is one that is common in games, but we do not do enough of this in our work processes. A key to making the measurement process effective is to make sure that the measurement is made as soon as the activity is completed. These measures should be oriented toward the measurement of the worth of the activity or process in relation to the goals and objectives.

The best person to make the measurement is the person performing the task. This allows that person to understand the measurement process, get feedback immediately, and offers the opportunity to be a stakeholder in the process.

The Cost of Quality

The cost of quality has several components. These components include the cost to do the job correctly every time, the cost of determining whether the output is acceptable, and all the other costs of the product not meeting either the specifications or expectations of the customer.

The costs of quality can be classified into several categories. The first of these costs are the direct costs of quality. Direct costs include costs that are discretionary such as prevention and appraisal costs.

Prevention costs include the costs associated with collection of data on quality and a reporting system for quality information. Other process control systems, which contribute to quality, also have costs associated with them. The costs of quality training and job-specific training are also prevention costs. The costs of vendor surveys to get information about products are prevention costs. The implementation costs are also considered prevention costs.

A second category of direct costs is the appraisal costs. These include typical items such as costs associated with external financial audits; approval signatures on documents; external endorsements from outside groups and agencies; the calibration of tests, apparatus, and methods; maintenance of testing equipment; design reviews; proofreading; and payroll audits.

Another major group of direct costs is termed resultant costs. These can be further categorized as internal and external resultants. Internal resultant costs might include such items as retyping letters, engineering changes, the extra costs due to late payment, and extra inventory costs. External resultant costs include such items as the cost of fixing or loss of downstream revenue of rejected products and services, as well as the costs of resolving differences, complaints, and dispute resolution. The administration costs of the warranty, the rework in the field or in the customer's shop, the extended and extra overhead for repair work, and the cost of missed opportunities due to the committing of limited resources also represent external resultant costs.

There are also indirect costs of quality that are harder to measure or even estimate, but exist and should be considered. These costs are typically on the customer's account initially, but ultimately become a cost to the contractor or have an impact on the contractor's operations. These costs include the loss of production by the customer, the extra time to discuss quality issues and problems, and the cost of rethinking the product expectations and needs.

Another area of indirect costs or impacts relates to the fact that quality problems can lead to customer dissatisfaction and ultimately loss of market share. These types of cost are difficult to quantify but ultimately the loss of reputation will result in a loss of opportunity.

Quantitative Measurement Tools

Quantitative measurement tools provide the quality improvement teams with objective information for making decisions about which processes to examine, where the problems might exist, and feedback concerning the impacts of changes in the processes. The quantitative measurement tools are used to provide benchmarks at all of the major customer interfaces, internal and external.

The quantitative measurements may represent data of two types. The data may be from published sources or be information obtained from original customer research about the firm's processes. When firms wish to identify benchmarks they can compare their performance with their current process, their major competition, or the "best practices." These data may come from generally available data that are published in trade magazines for some processes, from cooperative benchmarking studies of several participating firms, or from other generally available published sources. There are many examples of published cost and production rate tables that might represent these kinds of data.

For process improvement, the average values are not of much use and the best performance is what should be used as the benchmark. The typical approach for selecting an appropriate benchmark is to examine the entity that would provide the service or product if your firm did not. Thus, the office cleaning process should not be benchmarked against another contractor, but against another professional cleaning service.

Original research is the other way to establish benchmarks. There are several approaches and techniques to develop quantitative measurements. The first of these are interviews. The interview should be structured so that the information collected from several sources will be comparable if based on the same questions. Interviews can provide a rich set of data. The interview allows the responder to provide information without as much preparation as required in a formal written survey. Also, an interview allows the interviewer the opportunity to adjust the interview questions to explore in more depth areas where more information of interest is available.

Making site visits is an effective way to collect data concerning production processes and to identify the external environment constraints. Site visits can be used to learn more about your customer requirements, the integration needs, and the limits on your process. The site visit can also provide firsthand information on your own production processes. The best way to learn about a process is to watch it, record it, and interview the people involved in the process. This provides information not easily reduced to numbers, reasoning, and criteria concerning the process. The best way to understand the resource limits on a process, such as tools, materials, and information, is to watch the work being performed and to ask the craftsman what his or her limits are.

A more detached approach is to develop a survey to collect a set of data on a particular process or a product or service provided to an internal or external customer. The survey instrument typically is used after the problem has been bounded and deals with a specific set of issues related to a specific process. The survey must be carefully designed since there is not an opportunity to provide clarification or additional instructions orally. The survey questions must stand on their own and lead to single-ended results that are of interest to the surveyor.

It is important to carefully design a quantitative measurement tool such as a survey along several important dimensions, including scale, reliability, accuracy, and validity.

To be effective and useful, measurement tools should have a scale appropriate for diagnoses. In some cases, a simple yes or no may be the only required response. In other cases, it may be appropriate to provide a range of responses and guidance about the delineation between responses.

It may also be necessary to provide a calibration for the responses by using an example if the scale is one that respondents might be confused by. The survey instrument should also be reliable and measure the same thing, the same way, every time. This again requires some careful testing of the tool before using it on a wider scale. The tool also needs to be accurate for the purpose intended. This also carries over into the identification of the appropriate sampling technique and size to ensure that the tool provides the necessary delineation. Finally, the tool should represent a valid measure; it should measure what it is supposed to measure and not create more confusion. The purpose is to improve understanding, not create more uncertainty about what is happening.

Surveys can have a variety of types of questions, each with certain advantages and disadvantages for the type of information sought. The types of questions that are typically used are shown in Figure 8.17.

Customer Satisfaction

The key to knowing where to start the improvement process is to identify where there are problems or opportunities to increase customers' satisfaction. This step requires that you use a formal process to collect the information. A survey can

Multiple choice

Which of the following best represents the attitude of your supervisor?

(a) confident
(b) compassionate
(c) arrogant

Scale

On a scale of 1 to 10, with 10 being best, rate the company's accounting and payroll system.

Accounting system	1	2	3	4	5	6	7	8	9	10
Payroll system	1	2	3	4	5	6	7	8	9	10

Written comments

Describe any concerns that you have with our company's equipment maintenance.

Rating

Please rate (1 to 5) the following material suppliers on the following key areas;

XYZ electric distributor:

Price	1	2	3	4	5
Delivery	1	2	3	4	5
Service responsiveness	1	2	3	4	5

Forced choice

Which of the following best describes your foreman's attitude to safety? (circle one)

Unconcerned	Very concerned
Knowledgeable	Unknowledgeable

Figure 8.17 Survey question types

be a useful tool for this process. There are several types of surveys that can be developed:

- Owner surveys
- General contractor/construction manager surveys
- Employee surveys and suggestion boxes
- Supplier surveys

Each of these surveys provides a measure of the degree to which your processes are satisfying internal and external customers. Even your suppliers are your external customers in certain processes. For example, your suppliers can be thought of as external customers of your accounts payable department. These surveys do not have to be long documents. The surveys should be concise and directed toward the most critical requirements of your customer. Initially, the owner survey may focus on project objectives such as completing within budget and on schedule. In later stages of development, the owner survey may deal with process and interface issues. The same survey can be used over time to develop an understanding of the trends in performance and develop measures of improvement.

General contractor and construction manager surveys can be used by the specialty contractor to gauge the performance of the ongoing production processes and business processes as they relate to particular projects. These surveys, to be useful, should be initiated early and used monthly or more often with the progress billing cycle to provide timely feedback and guide the improvement process while the project can still benefit from the changes. Since the relationship with the designer is not a direct one, input concerning such items as shop drawing review should be addressed through this customer who ultimately will provide feedback from the true customer.

Internal customers are employees who have enormous potential to improve the firm's processes. This storehouse of knowledge, expertise, and experience can be tapped through an internal customer survey. This survey may start out in a general way to explore where the problems and opportunities exist internally and gradually evolve to surveys focused on specific internal processes, such as timekeeping. These surveys should be repeated periodically to gauge positive impact on the organization from improvement projects undertaken.

Suppliers should also be surveyed to learn where the interfaces between the two organizations can be made to work better. Wherever there is an exchange, there is the combination customer-supplier relationship. This process also establishes a communications channel and provides the opportunity for the supplier to better serve their customer, you.

The surveys that are developed and used will vary widely between companies and over time. There is not one solution to the survey design. Basic principles of survey design should be followed. Questions should be worded to avoid establishing a biased answer. The opportunity for open-ended feedback should be provided along with fairly, narrowly defined subjects. Finally, the survey should not unduly burden the responder in terms of the time required for completion. It is usually appropriate to limit the survey to something that can be completed within less than 20 minutes.

CONCLUSIONS

This chapter presents a somewhat generic approach to a TQM program. Each company will necessarily customize the approach to fit its own culture, level of commitment, point in the understanding process, and level of financial resources. In addition, the model TQM program presented here may be adjusted up or down depending on the size and needs of the organization. However, the steps in the implementation process should not change and the time required for seeing benefits may vary substantially from company to company.

Finally, to successfully implement TQM, you must learn about these principles in greater detail. We suggest the creation of your own library of TQM-related books and articles to enhance your understanding of TQM and the specific knowledge required for implementation. These books and articles could also be provided to key personnel within your firm to begin the awareness process.

There are literally hundreds of books and articles published on quality and continuous improvement. Every day, more and more of these articles and books appear. We recommend the following as a core list: *Thriving on Chaos: Handbook for a Management Revolution* by Tom Peters (1987); *The Deming Management Method* by Mary Walton (1986); *Juran on Leadership for Quality* by Joseph Juran (1989); "Successful Change Programs Begin with Results" by Robert Schaffer and Harvey Thomson (1992); *Methods Improvement for Construction Managers* by Henry Parker and Clarkson Oglesby (1972); and *The Team Handbook* by Barbara Streibel, Brian Joiner, and Peter Scholtes (2003). Complete references are provided in the Bibliography section.

WEB ADDED VALUE

AssessingTQM.pdf

This file provides a methodology for assessing contractors' TQM program progress based on the Malcolm Baldrige National Quality Award (Baldrige) Criteria. The Baldrige criteria represent an effective diagnostic tool that can be used by the contracting firm for self-assessment and planning. The results of this self-assessment can then be used to adjust the contracting firm's direction to meet the competitive challenges of today and tomorrow.

Web
Added
Value™

This book has free material available for download from the
Web Added Value™ resource center at *www.jrosspub.com*

9

QUALITY ASSURANCE

Dr. Thomas E. Glavinich, *University of Kansas*

INTRODUCTION

The purpose of this chapter is to provide guidance in the preparation and implementation of quality assurance (QA) programs for construction. This chapter provides construction firms with the means to easily develop a customized QA program that will improve the quality of construction and increase productivity. The QA program should also serve as a feedback mechanism for the continuous improvement of field operations.

Successful construction firms already have effective QA programs in place or they would not be successful. What is usually missing is the formalization of this program in writing. In most instances, the development of a formal QA program will involve only documenting those processes that already exist within the construction firm. The documenting of existing operations should lead to a better understanding of the firm's processes as well as their improvement.

This chapter provides a comprehensive starting point that can be customized and adapted for the contractor's own organization and needs. By following the advice in this chapter, contractors should be able to expedite the program development process and save a great deal of time and effort.

The ISO 9000 series of quality management and assurance standards was chosen as the basis for the material presented in this chapter. This series of standards is promulgated by the International Organization for Standardization (ISO) in Geneva, Switzerland. The ISO 9000 standards are the internationally recognized standards for QA in the manufacturing and service industries. These standards have been adopted by countries throughout the world and are becoming the basis for international trade in goods and services. In the United States, these standards have been adopted by the American National Standards Institute (ANSI) through the American Society for Quality (ASQ).

QUALITY ASSURANCE FUNDAMENTALS

Quality assurance (QA) is a broad term that refers to the development and application of procedures that ensure that a product or service meets the customer's performance criteria. In construction the customer's performance criteria are defined by the project contract documents, which include the plans and specifications. QA is concerned with ensuring compliance with the contract documents through the systematic monitoring and control of the construction process on an ongoing basis. The goal of QA is that the finished work in place is in compliance with the contract documents avoiding costly and time-consuming rework to correct deficiencies.

It is important to note that QA is not about continuous improvement. Continuous improvement is the domain of Total Quality Management (TQM). QA is ensuring that the processes are in place to guarantee that the customer receives what is needed when it is needed. QA is about giving the customer confidence that your firm can meet his or her needs and expectations per the contract documents. Customer confidence comes from having key construction processes documented that will ensure all necessary steps and procedures will be satisfactorily completed when installing and testing materials and equipment.

The terms quality assurance (QA) and quality control (QC) are often used interchangeably. QA and QC are not the same and this chapter differentiates between them. QC is just one part of the overall QA program. Simply stated, QA is proactive with regard to project quality whereas QC is reactive.

QC is concerned with checking that the completed work in place is in compliance with the project plans and specifications. Any identified deviation from the plans and specifications is reviewed, and if warranted, action to correct the deficiency is taken. QC focuses on correcting deviations from the plans and specifications once the required work is in place. QC leads to an expensive correction of quality problems, which could be avoided through an effective QA program.

An example of a common QC activity in the construction industry is the "punch list." The punch list is important because it gives the owner and architect/ engineer an opportunity to review the completed work and determine if it is in compliance with the construction documents prior to accepting it. The punch list should not be used as a substitute for an effective QA program.

In Chapter 8, we reviewed Dr. Deming's "14 Points for Management." Point 3 advises management to cease dependence on inspection to achieve quality. According to Deming (1986), reliance on inspection is the same as planning for defects. By relying on inspection to improve quality, the contractor is acknowledging that the construction process is incapable of conforming to plans and specifications the first time and rework will always be required. Inspection neither improves nor guarantees quality. It is not possible for contractors to inspect quality into the completed work.

A successful project is a project that is completed in accordance with the contract documents, on schedule, and within budget. Contractors need a QA program to ensure that the completed project meets the project requirements as stated in the contract documents. An effective QA program will help the construction firm achieve a successful project. A QA program will improve quality and help prevent problems caused by oversight and lack of attention to detail.

All successful construction firms have a QA program. This QA program may be informal and not in writing, but everyone associated with the firm knows that it exists. The management of the construction firm conveys its dedication to quality and customer satisfaction through both its words and actions. However, on many projects today contractors are being required by the owner to provide a written QA program as part of the prequalification process or prior to start of work. This is especially true for those owners who have an active commitment to formal QA programs and have either attained or are working toward ISO 9000 registration.

There is increasing external pressure from quality conscious owners for the construction firm to formalize its QA program and commit it to writing. Formalizing the QA program provides advantages to the construction firm beyond the fulfillment of a project requirement. A formal QA program will benefit the entire organization as well as other projects not requiring a formal QA program. The very act of preparing a formal QA program requires that the contractor work through and document all of the processes that affect construction quality. Once these processes are documented, the information contained in the formal QA program can be used as a baseline for process improvement as well as a tool for communicating the firm's commitment to quality both within and outside the firm.

Quality Assurance Program Costs

The costs associated with the development and implementation of a formal QA program for the construction firm include:

- **Original Quality Assurance Program Development.** The cost of the original QA program development can be substantial for the construction firm. This is because key employees must be involved in the development of the QA program. These key employees will need to be trained in QA and then will need to develop the QA program. All of this takes these key employees away from their daily work, which means someone else will have to perform their work or the work will not be completed. The use of outside consultants may help to facilitate the program development process, but consultants cannot be substituted for the involvement of the firm's key employees. To be effective and successful, the QA program must fit within the firm's existing culture and employees must feel a sense of ownership with regard to the program.

- **Ongoing Quality Assurance Program Evaluation and Upgrading.** Once the QA program has been developed, there must be an ongoing evaluation and upgrading of the program. This again requires the involvement of key employees throughout the firm and a substantial investment of their time. If the QA program is not continuously evaluated and upgraded, it will soon become outdated and useless.
- **Initial and Ongoing Employee Quality Assurance Training.** The QA program will not work without the involvement of all the employees working for the construction firm. The firm's employees, no matter what level, must be made solely responsible for the quality of their own work. Along with this responsibility must come empowerment or the authority to take corrective action when the employee sees the need to ensure the quality of their work. Without employee participation, QA will regress back to QC with management unsuccessfully attempting to inspect quality into the finished work.

All employees must receive initial training to acquaint them with the firm's QA program. In addition, all employees must receive ongoing training where appropriate regarding changes to the QA program. All of this takes employee time and can be expensive. However, there is no substitute for employee involvement.

Quality Assurance Program Benefits

The benefits of an effective QA program should far outweigh the costs. In fact, the cost of developing and implementing the QA program should be seen as an investment in the firm's future. Benefits from the successful implementation of a QA program should include higher employee morale, greater productivity, less rework, avoidance of claims, and satisfied customers. Individually, these benefits may be difficult if not impossible to quantify on individual projects. However, as a whole these benefits should manifest themselves in increased market share and greater profits for the firm.

Connection between Quality Assurance and Total Quality Management

The QA program should be an extension of the construction firm's in-house TQM program. The QA program should address the contractor's production processes, which are critical to the survival and continued growth of the firm. An effective QA program will provide a valuable feedback mechanism for continuous improvement of construction processes.

ISO 9000

The ISO 9000 series standards are the internationally recognized standards for QA in both manufacturing and service industries. These standards have been adopted by countries throughout the world and are becoming the basis for international trade in goods and services. In the United States, these standards have been adopted by the American National Standards Institute (ANSI) through the American Society for Quality (ASQ) as the ANSI/ASQ Q9000 series of quality management and QA standards that will be referenced throughout this chapter. The ANSI/ASQ Q9000 standard series is equivalent to the ISO 9000 standard series.

By basing your QA program on the ANSI/ASQ Q9000 standards, the QA program should satisfy owner requests for information about the construction firm's QA program. Additionally, the preparation of a QA program in accordance with the ANSI/ASQ Q9000 standards can serve as a first step toward ISO 9000 registration.

There are five individual standards that make up the ANSI/ASQ Q9000 standard series and each of these standards has an ISO standard counterpart. ANSI/ASQ Standard Q9000 presents the principal quality concepts and lays the groundwork for ANSI/ASQ Standards Q9001, Q9002, and Q9003. ANSI/ASQ Standard Q9001 is the most comprehensive standard of the group and deals with the design, development, production, installation, and servicing of the product or service. ANSI/ASQ Standard Q9002 is more restrictive in its scope excluding design and development. ANSI/ASQ Standard Q9002 deals only with the production, installation, and servicing of the product or service. Even more focused is ANSI/ASQ Standard Q9003, which deals only with the final inspection and testing of the product or service. ANSI/ASQ Standard Q9004 provides guidance regarding the design and implementation of quality systems within the firm itself.

Typically, contractors tend to base their QA programs on ANSI/ASQ Standard Q9002 because this standard deals with the installation of a product in accordance with given plans and specifications. However, this chapter is based on the more comprehensive ANSI/ASQ Standard Q9001, which includes design and development. The rationale for this is:

- Many contractors produce custom equipment for customers based on performance specifications

- There is a trend in the construction industry toward more design-build work
- Contractors have increasing responsibilities to their customers for guarantees and warranties, ongoing maintenance, and servicing of systems after installation

QUALITY ASSURANCE PROGRAM MODEL

Both ANSI/ASQ Q9001 and Q9002 include a Section 4, which is broken down into 20 paragraphs that provide quality system requirements for conformance with these standards. Both standards are intentionally written generically so that they can be applied to any manufacturing or service industry. This chapter adapts the ANSI/ASQ Q9001 and Q9002 requirements to the construction industry.

Table 9.1 provides a listing of the ANSI/ASQ Q9001 and Q9002 sections. As can be seen from the table, both ANSI/ASQ Q9001 and Q9002 contain the same sections. The major difference between these two standards is in Paragraph 4.4/Design Control. ANSI/ASQ Q9001 provides detailed quality system requirements for design control in Paragraph 4.4. Paragraph 4.4 of ANSI/ASQ Q9002 states that this standard does not cover design control and that this section is included only to keep the section numbers consistent with ANSI/ASQ Q9001. Please refer to Table 9.1 when reviewing the following sections to identify the corresponding ANSI/ASQ Q9001 and Q9002 paragraphs.

Management Responsibility

Management responsibility is divided into the following three requirements:

- **Quality Policy.** Contractors must define and document their quality policy and make an overt commitment to quality. ANSI/ISO/ASQ A8402 defines "quality policy" as the overall intention and direction of an organization regarding quality as formally expressed by management.
- **Organization.** Contractors are required to define the responsibility of all personnel with regard to quality. This includes the identification of personnel responsible for verifying that the work in place meets specified requirements. Additionally, contractors must assign the responsibility for direct oversight of the quality program to one member of the contracting firm's upper management.
- **Management Review.** The QA program must be reviewed by the contractor on a regular basis to ensure its continued validity and effectiveness.

Table 9.1 Quality system requirements

ANSI/ASQ Q9001 and Q9002 Quality System Requirements	ANSI/ASQ Standard	
	Q9001	Q9002
Management responsibility	4.1	4.1
Quality system	4.2	4.2
Contract review	4.3	4.3
Design control	4.4	4.4
Document and data control	4.5	4.5
Purchasing	4.6	4.6
Control of purchaser-supplied product	4.7	4.7
Product identification and traceability	4.8	4.8
Process control	4.9	4.9
Inspection and testing	4.10	4.10
Control of inspections, measuring, and test equipment	4.11	4.11
Inspection and test status	4.12	4.12
Control of nonconforming products	4.13	4.13
Corrective and preventive action	4.14	4.14
Handling, storage, packaging, preservation, and delivery	4.15	4.15
Control of quality records	4.16	4.16
Internal quality audits	4.17	4.17
Training	4.18	4.18
Servicing	4.19	4.19
Statistical techniques	4.20	4.20

Quality System

"Quality system" is defined in ANSI/ISO/ASQ 8402 as the organizational structure, responsibilities, procedures, processes, and resources needed to implement the QA program, which is aimed at achieving the quality objectives. The construction firm must develop and maintain a documented QA program in the form of a QA manual and implement that program.

Contract Review

ANSI/ASQ Q9001 and Q9002 require that the contractor develop procedures for contract review to ensure that:

- The requirements of the contract for providing construction services are adequately defined and documented

- Any contractual requirements that differ from the bid documents are identified and resolved
- The construction firm has the resources and expertise to meet the contract requirements

A record of the contract review and any subsequent discussions with the client concerning the contract must be maintained by the contractor.

Design Control

When the project requires that the contractor provide design services as in the case of a design-build project, contractors are required by ANSI/ASQ Q9001 to ensure that the project owner's needs and requirements are met. Specifically, ANSI/ASQ Q9001 requires that the contractor perform the following activities:

- **Plan and Schedule the Design Process.** The contractor undertaking a design-build project must first plan and schedule the design process. The planning of the design process must involve the identification of:
  - Responsibility for supplying or obtaining design information.
  - Systems that are to be designed and who is responsible for those systems. For example, the design of a life safety/security system for a commercial building might be the responsibility of the manufacturer of the selected system.
  - Design milestones that must be met. For example, specific dates when design documents must be available for construction to proceed as in the case of fast-track construction projects.
  - Interface points between various subsystems that the contractor is responsible for designing as well as interface points with other systems that are not the responsibility of the contractor.
  - Design review points to ensure that the client's needs and requirements are met by the systems designed.
- **Identify Organizational and Technical Interfaces.** Contractors need to identify organizational and technical interfaces to facilitate the design process. These organizational and technical interfaces exist both within the contracting firm and with outside entities that have a stake in the design. These outside entities include the owner, other specialty contractors, insurance carriers, material and equipment manufacturers and suppliers, specialty design consultants, code authorities, and others.

  Contractors must identify individuals within the owner's organization that have the ability to make design decisions as well as obtain design criteria and other needed information from the end user of the system or facility. Additionally, the contractor needs to identify individuals within the material and equipment suppliers' organization who will be responsible for the project and are capable of providing technical information as required and committing to design and delivery schedules. This information should be documented and distributed to everyone needing to know these contact points.

- **Assign Design Activities.** The responsibility for the performance of the various design activities needs to be assigned and documented. In the case of joint ventures with design firms and arrangements with manufacturers and suppliers providing system design, this documentation would take the form of the scope of services included in the contract documents. For in-house personnel, written work assignments should be developed, negotiated, and agreed to by individuals responsible for the work. In either case, this documentation should clearly state the scope of work or systems to be designed, schedule for design, and delineate the design deliverables, and any other nontechnical design requirements that need to be addressed.

- **Determine the Owner's Design Requirements.** The design team must identify and document the owner's design requirements to ensure that the completed project meets the owner's needs and requirements. The owner's performance specification must be converted into specific and measurable design requirements and reviewed with the owner to ensure accuracy and completeness. Any conflicting or ambiguous requirements must be resolved prior to proceeding with the design.

- **Document the Design.** The design must be documented as it proceeds. Documentation certainly includes the plans and specifications that will be used to order materials and equipment from manufacturers and suppliers as well as put work in place in the field. However, design documentation also includes information gathered about existing conditions at the physical site, manufacturer and supplier information, design calculations, code reviews, and design memoranda, among other relevant information.

- **Perform Design Reviews.** Regular design reviews should be planned at critical points during the design process. These design reviews should involve the owner and design team as a minimum. In addition to ensuring that the evolving design meets the owner's technical needs and requirements, these design reviews should also include constructability reviews and value analyses to ensure that the project can be efficiently and economically built and operated as designed.

- **Address Design Changes.** The construction firm must have procedures in place to identify, document, and approve changes in the design when they occur for any reason.

Document and Data Control

ANSI/ASQ Q9001 and Q9002 require that the contractor develop and implement procedures for project document control. For design-build projects, this document control requirement includes the cataloguing and filing of all design documents as well as construction documents. The purpose of this requirement is to ensure that all the project requirements are met and documented for future reference.

During construction, the contractor must have procedures in place that ensure current construction documents are available in the field. These procedures should make sure that the latest revision of the drawings and specifications are available, all field directives and change orders are noted on the latest revision of drawings and specifications, and all superseded documents are removed, catalogued, and filed for future reference. In addition, document control includes documenting daily field operations and the filing of all correspondence, memoranda, schedules, budgets, inspection and testing reports, shop drawings and catalogue cuts, and material and equipment shipping and receipt records, among other documents.

In addition to maintaining project documents for use in managing the project's construction, the contractor must also ensure that field employees have the information needed to build the project in accordance with specified requirements. This requirement can be met by providing ready access to the latest construction documents and carefully planning day-to-day construction operations at the site.

Purchasing

It is the contractor's responsibility to ensure that materials and equipment purchased for incorporation into the project meet the technical requirements specified in the plans and specifications. Per ANSI/ASQ Q9001 and Q9002, ensuring that purchased materials and equipment meet specified requirements involves the following processes:

- **Assess Suppliers.** Suppliers should be selected based on their ability to supply materials and equipment that meet the project's technical requirements and delivery schedule. In addition, the supplier's commitment to quality, QA program, and the contractor's past experience with the supplier should be part of the selection criteria. The supplier selection criteria and the selection process should be documented and followed. If a single-source supply is specified, the contractor should assess the supplier's capabilities and establish a working relationship as soon as possible to avoid problems and miscommunication during construction.

- **Review Purchasing Data.** Prior to purchase, ANSI/ASQ Q9001 and Q9002 require that the contractor request and review purchasing data to ensure that the material and equipment ordered will meet the technical requirements of the project. Contractors should verify that the purchased material and equipment will meet the project requirements through catalogue cuts, shop drawings, samples, certified testing results, or visits to supplier production and/or testing facilities.

Where the specification requires that the contractor submit purchasing data to the owner and/or engineer for review and approval, contractors should review the data in advance to ensure that they meet both the specified technical and submittal requirements prior to forwarding the data on to the general contractor, owner's representative, and/or engineer. A log of all purchasing data required, submitted, and the status of that information in the review process should be maintained for the project as part of the project document control requirements. A log such as this is commonly kept during construction and referred to as a "shop drawing log." A copy of all purchasing information should be kept on file for ready reference during construction.

Control of Purchaser-Supplied Product

ANSI/ASQ Q9001 and Q9002 require that the contractor establish and maintain procedures for the verification and storage of owner-furnished equipment and materials.

Product Identification and Traceability

The contractor should maintain records of the installation of materials and equipment. This requirement of ANSI/ASQ Q9001 and Q9002 can be met by establishing procedures in the field that ensure an accurate and complete set of as-built record drawings are kept. In addition, records of where specific materials and equipment are installed should also be kept. For example, if similar equipment from multiple manufacturers is used on a particular project, knowing which manufacturer's equipment is installed where may be helpful during system start-up and troubleshooting throughout the life of the installation.

Process Control

The production process in construction is the installation of materials and equipment at the project site. ANSI/ASQ Q9001 and Q9002 require that the installation of materials and equipment at the site be carried out under controlled conditions. To control the installation process, the contractor must do the following:

- **Pre-plan the Work.** Work in the field at the activity level must be planned in advance to ensure quality. Employees must have the technical information, skills, tools and production equipment, materials and equipment, and place and time in order to perform the work correctly.

    Without any one of these ingredients, quality and productivity will suffer.

Pre-planning a construction activity documents work instructions and defines the means and methods of installation. Pre-planning goes beyond the end result that is defined in the plans and specifications. Pre-planning works out in advance how the end result defined in the plans and specifications can be accomplished. Pre-planning defines the construction process at the activity level.

Pre-planning starts with defining the scope of a construction activity based on the construction plans and specifications. Construction activities should be defined by the contractor's overall project schedule and budget. Once defined, the means and methods for accomplishing the construction activity must be identified and documented. Wherever possible, the means and methods should either be worked out in conjunction with the employees performing the work or reviewed with the employees prior to finalizing the pre-plan. In addition to defining what the work is and how it will be accomplished, other information that should be part of the pre-plan is as follows:

- Drawing and specification references as well as any additional activity requirements or restrictions contained in the general, supplemental, or special conditions
- Crew size and skill mix allotted to the activity
- Type, quantity, and location of materials and equipment, including expendables
- Type, quantity, and location of tools and production equipment, including those supplied by other entities on-site such as cranes for lifting and hoisting
- Location(s) of the work to be performed
- Time(s) allotted to perform the work
- Interface points with the owner, engineer, manufacturers and suppliers, other trades, or other outside entities
- Budget for performing the work defined in dollars, employee hours, or material quantities if appropriate
- Where the crew goes once this work is finished

Monitor and Control the Work

During the performance of the work, it is the contractor's responsibility to monitor and control the quality and progress of the work. Monitoring and controlling the work can be accomplished in a number of different ways, which include observation and regular meetings with construction crews. Wherever possible, quality and progress should be tracked quantitatively, preferably graphically, and shared with the crews. In this way, potential problems in the construction process can quickly be identified and resolved with the help of the crew performing the work.

Test and Inspect the Work

Tests and inspections should be carried out during construction wherever required by the technical specifications or the contractor's QA program. These tests and inspections should be documented as to their results and the planned corrective action that will be taken when necessary. Finally, the resolution of the problem or the results of the corrective action should also be documented.

Establish Criteria for Workmanship

The criteria for judging workmanship must be defined by either the technical specifications or the contractor's QA program. Workmanship criteria should be reasonable, achievable, and measurable and must be made known to the employees performing the work.

Workmanship criteria should be documented and, wherever possible, reference industry codes and standards.

Inspection and Testing

Contractors must maintain records of all inspection and testing that includes the results of those inspections and tests. Contractors are required, by ISO 9001, to test and inspect materials and equipment as follows:

- **Receiving Inspection and Testing.** Contractors must ensure that incoming materials and equipment are not incorporated into the work until they have been inspected and verified as conforming to the project requirements. Verification should be in accordance with either the technical specifications or the contractor's QA program. It is the contractor's responsibility to determine how much inspection of incoming materials and equipment needs to be completed given the supplier's QA program and past experience with the supplier.
- **In-Process Inspection and Testing.** Contractors must inspect and test work in process where required by either the project technical

specifications or the contractor's QA program. Documented inspection and test methods must be used to determine whether or not the work in process is in conformance. Wherever possible, criteria for conformance should be quantifiable and measurable. Procedures also need to be in place to identify nonconforming work and initiate corrective action as soon as possible.

- **Final Inspection and Testing.** ANSI/ASQ Q9001 and Q9002 require that contractors perform final inspection and testing in accordance with the project technical specifications or the contractor's QA program prior to turning the completed work over to the owner. The final inspection and testing requirement also states that the contractor must ensure that all required receiving and in-process inspection and testing have also been performed. Records of final inspection and testing should be kept.

In construction, the final inspection and testing is usually considered to be the "punch list" prepared by the owner and/or engineer following a walk-through of the completed project. Under ANSI/ASQ Q9001 and Q9002, it is not sufficient for contractors to use the "punch list" as a measure of quality. Contractors must take a proactive role and perform all necessary inspection and testing prior to the final inspection and testing by the owner and/or engineer. It should be the contractor's goal to have all problems identified and, if possible, corrected prior to the final inspection and testing by the owner and/or engineer.

Control of Inspections, Measuring, and Test Equipment

Contractors are responsible for calibrating and maintaining test equipment used on the project whether it is owned, leased, or borrowed. In addition, only equipment suitable for the required inspections, measurements, and tests should be used. Equipment must be used so that measurements are made within the equipment's capabilities. The following guidelines regarding the use of equipment should be adhered to:

- Measurements should be taken in accordance with material and equipment manufacturer's recommendations and industry standards
- Equipment should be appropriate for the application and capable of the measuring accuracy and precision necessary
- Calibration of equipment should be performed at prescribed intervals or prior to use
- Equipment should be clearly marked to show its calibration status
- Calibration and maintenance records should be maintained for equipment

- Environmental conditions should be suitable for the calibration and use of the equipment
- Equipment should be handled, transported, and stored so that its accuracy and calibration is maintained
- Where auxiliary test hardware or software is used with inspection, measuring, and test equipment, it should be checked on a regular basis for accuracy and suitability for use

Inspection and Test Status

The inspection and test status of materials, equipment, and work in place must be identified by the contractor. Inspection and test status includes both conformance and nonconformance with the inspection and test criteria. ANSI/ASQ Q9001 and Q9002 require that the inspection and test status of materials, equipment, and work in place be clearly identified using markings, stamps, tags, labels, inspection records, software, physical location, or other suitable means. The required inspection and test status indication must remain in place throughout construction.

Control of Nonconforming Products

Contractors must establish and maintain procedures to ensure that materials, equipment, or work in place that does not conform to specified requirements are identified and corrected in order to comply with either ANSI/ASQ Q9001 or Q9002. Nonconforming materials, equipment, and work in place must be rectified in one of the following three ways:

1. Reworked or modified to meet specified requirements
2. Accepted with or without rework or modification by concession
3. Removed and replaced in total

Where nonconforming materials, equipment, or work in place is accepted as is, the contractor must document the nonconformance and that it has been accepted. Replaced, reworked, or modified materials, equipment, or work in place must be reinspected and retested in accordance with the technical specifications and/or the contractor's QA program.

Corrective and Preventive Action

To comply with the requirements of ANSI/ASQ Q9001 and Q9002, contractors must develop, document, and implement procedures to do the following:

- Investigate and analyze the cause of the nonconformance
- Define the corrective action necessary to prevent recurrence

- Initiate corrective action
- Ensure that the corrective action taken is effective
- Revise existing procedures and processes in accordance with the corrective action taken

Handling, Storage, Packaging, Preservation, and Delivery

Procedures must be developed, documented, and implemented by contractors for doing the following:

- **Handling Materials and Equipment.** Contractors must develop, document, and use means and methods of handling materials and equipment that prevent damage or deterioration.
- **Storing Materials and Equipment.** Contractors must provide secure storage areas to prevent damage or deterioration of the materials and equipment prior to integration into the work. In the case of bulk materials, contractors should establish material control procedures to maintain the integrity of the inventory at the site.
- **Packing Materials and Equipment.** Materials and equipment must be properly packed for movement. This includes informing the manufacturers or suppliers as to the anticipated conditions that materials and equipment will encounter during shipment. This requirement is especially important when materials and equipment will be temporarily stored at an intermediate location prior to being shipped to the project site.
- **Delivering Materials and Equipment.** ANSI/ASQ Q9001 and Q9002 require that contractors protect materials and equipment from damage during delivery.

Control of Quality Records

To comply with ANSI/ASQ Q9001 and Q9002, contractors must maintain quality records to demonstrate that the specified level of quality has been attained. Contractors must establish and maintain procedures for identifying, collecting, cataloguing, and filing these records. Supplier and subcontractor quality records should also be a part of the contractor's quality records.

Internal Quality Audits

ANSI/ASQ Q9001 and Q9002 require that contractors establish, document, and carry out a comprehensive system of internal quality audits. The purpose of these internal quality audits is to ensure compliance with the QA program. The internal quality audits should be performed in accordance with documented procedures. The results of the internal quality audits should be documented and shared with the personnel who are responsible for the processes being audited.

Training

Procedures must be developed, documented, and implemented to ensure that all employees performing activities that affect quality are qualified. Employees can be qualified on the basis of education, training, and/or experience. Where training is required for employees to perform a particular task, it is the contractor's responsibility to ensure that effective training is provided. Contractors must maintain training records.

Servicing

When after-construction servicing is part of the construction contract, contractors must establish, document, and implement procedures to ensure that the servicing meets stated requirements. The stated requirements include those specified in the technical specifications as well as those required by the contractor's own QA program. For example, it is common in many construction contracts to require the contractor to warrant the installation for one year after substantial completion. In addition, there are manufacturer and supplier guarantees and warranties that must be honored.

Statistical Techniques

Where the contractor uses statistical techniques to verify the acceptability and conformity of materials, equipment, or work in place, the statistical techniques must be appropriate for the application and properly applied.

STEPS IN DEVELOPING A QUALITY ASSURANCE PROGRAM

Step 1: Establish a Quality Assurance Program Team

The first step in developing an effective QA program is the establishment of an in-house cross-functional team that is responsible for the program development. Members of this team must have an in-depth knowledge of the firm, its operations, and an understanding of QA. The QA program team should report directly to upper management through the member of the management team who has been assigned the responsibility for quality within the construction firm.

QA program team members must be carefully selected. Team members should come from the various functional areas within the firm that will be addressed in the firm's QA program. It is not necessary that there be a team member representing each function covered in the firm's QA program. It is only necessary that the team as a whole be knowledgeable in all the processes that will be addressed by the QA program. The size of the team will, of course, depend on the size of the construction firm and the complexity of its internal operations and the projects it takes on. The QA program team should not exceed four to six members, including the member from upper management.

The member of upper management who is responsible for the firm's QA program should be a member of the QA program team. The reason is that this individual will be responsible for leading the implementation effort and must have a thorough knowledge of how the QA program was developed. In addition, active participation of a management team member demonstrates upper management's commitment to quality and the QA program.

Formation of the QA program team and its efforts in the development of an effective QA program is the first opportunity upper management has to demonstrate its unwavering support for the program and its development. Normally, members of the QA program team will be the firm's best performers and current or future leaders. These individuals will necessarily be in middle-management positions, which will provide them with the knowledge of how the firm actually operates on a day-to-day basis.

Being part of middle management, these individuals have key responsibilities for ongoing operations within the firm. These responsibilities cannot be neglected while developing the firm's QA program. To put these individuals in the position of neglecting ongoing operations in the interest of developing a QA program would be a mistake. These individuals would rightly forsake the development of the QA program to take care of their day-to-day responsibilities. This would doom the QA program to failure before it even got started.

It is here that management must provide these individuals with the additional support they need to continue to meet their day-to-day obligations as well as have the time needed to develop the QA program. The additional support can take a variety of forms, including the temporary assignment of existing personnel, redefinition of subordinate duties and responsibilities, or bringing in outside help on a temporary basis.

Some middle managers may not be comfortable delegating key responsibilities to subordinates. If this is the case, then upper management may want to reconsider the selection of that individual. To have an effective organization, there must be trust at all levels and a dedication to employee growth and development. A middle manager who will not delegate, for whatever reason, is stifling the firm's growth and either needs additional training or replacement.

Once the QA program team has been established, upper management must draft and provide the team with a clear and unambiguous mission statement. This mission statement must state what is expected of the team and the scope of their effort. This mission statement should also note available resources, time frames, and deadlines as well as any restrictions on the QA program team's operations.

Step 2: Develop a Quality Assurance Plan

The first thing that the QA program team should tackle is the development of the QA implementation plan. Within the confines of the mission given to the QA program team by upper management, the QA program team should decide exactly how it is going to achieve the stated objectives. This plan will detail how the QA program will be developed, who will be responsible for what, when things should be completed, and what the final product will include.

Step 3: Assign Quality Assurance Responsibility

Within the framework of the organization, responsibility for quality needs to be established. Everyone in the organization is responsible for the quality of his or her own work. However, certain people within the organization must have the overall responsibility for the implementation of the firm's QA program. In particular, these people are responsible for ensuring that the QA procedures developed and documented as part of the QA program are carried out as well as the training of people within the organization in these procedures.

Assignment of responsibility for QA should be accomplished by job title only and not by individual. The assignment should be made based on the position's responsibilities, authority, and accountability. It is important that responsibility for QA be established early because the position responsible for a particular portion of the QA program may determine the procedures developed.

The assignment of QA responsibility may also require the revision of job descriptions throughout the organization. Everyone will now be responsible for the quality of his or her own work. To ensure that everyone understands their role in the firm's quality journey, employee job descriptions should be expanded to reflect each individual's responsibility for quality.

Step 4: Identify and Analyze Processes

Once responsibility for QA has been determined, the next step is to identify and analyze the processes that will form the basis for the firm's QA program. The processes selected must all contribute to quality and will vary to some extent from contractor to contractor. However, these processes will probably be the same for most contractors serving the same market. Typical construction processes include the following:

- Contract document review
- Document control

- Design management
- Procurement and expediting
- Tool and equipment management and calibration
- Material and installed equipment management
- Construction management
- Inspection, testing, and start-up

The QA program team does not get into the details of continuous improvement of these processes. The QA program team's objective is to document existing processes. If improvement is required, the improvement of existing processes should be left to specially formed quality improvement teams as part of the firm's overall continuous improvement or TQM program. The QA program team cannot get involved in the improvement of processes or it will never complete its work.

Step 5: Establish and Document Procedures

Once the construction firm's processes have been identified and analyzed, the next step is to establish and document procedures used to carry out those processes. Procedures can be documented in a variety of ways that include anything from verbal descriptions to graphical representations or a combination of methods. The key is to provide enough detail in the process description so that anyone with minimal training can understand and effectively carry out the process. One test often applied to determine the adequacy of process documentation is if all personnel were suddenly replaced, could the new people with proper training use the process documentation to continue providing the product or service as before.

Step 6: Write the Quality Assurance Manual

The cornerstone of the QA program is the QA manual. The QA manual provides an overall guide that describes how the construction firm's QA system works. Where additional detail is required, the QA manual references more detailed documents for specifics.

Step 7: Provide Quality Assurance Training

The QA program team that develops the QA program cannot be charged with its implementation. Implementation of the QA program must involve everyone working for the construction firm. Therefore, contractors must provide training to employees regarding what the QA program is and their role in its implementation.

Step 8: Implement the Quality Assurance Program

The last step in the development of an effective QA program is its implementation. The QA program must be implemented across the organization in accordance with the QA manual.

The QA program should be dynamic and not static. The QA program should be reviewed on a regular basis and improved constantly. It is not enough to simply develop a QA program. The processes on which the QA program is built should be in a state of continuous improvement. This means that the QA program will need to be constantly updated so that it accurately reflects the contractor's current processes.

INSTALLATION, INSPECTION, AND TESTING PROCEDURES

The establishment and documentation of procedures is Step 5 in the development of an effective QA program as explained earlier. These documented procedures should be developed by the contracting firm for both its business and production processes. These documented procedures form the basis of a procedure manual that supports the contractor's QA program.

These procedures support and are an integral part of the contractor's QA program. The goal of these procedures is to minimize variance in the completed work that will result in higher quality, reduced rework, and improved productivity.

The documentation of the contractor's production processes directly affects the quality of work in place and can benefit the greatest from standardization. Business processes are also important and should also be documented, but these processes are unique to each contracting firm and depend greatly on the organization structure, size of the firm, market served, and personalities involved. This section focuses on the installation, inspection, and testing procedures that directly affect the quality of work in the field which is what interests the customer the most.

Documenting Production Procedures

Procedures for the installation, inspection, and testing of materials, equipment, and systems must be developed to ensure that the completed work meets the customer's technical requirements as expressed in the contract documents. Documented procedures support the QA program by providing employees with detailed information on how work should be carried out. By documenting these procedures, contractors ensure that installation, inspection, and testing are carried out consistently throughout each project. In addition, having written procedures allows contractors to compare their procedures with those required by the project technical specifications or the manufacturer's requirements to ensure contract compliance.

Production Procedure Basis

The contractor's documented production procedures should be based on the following:

- **Project Technical Specifications.** The project technical specifications along with the general, supplemental, and special conditions; drawings; and other documents referenced in the agreement between the contractor and customer are part of the contract documents. These contract documents establish the customer's requirements for the project, which objectively define quality. Therefore, the contract documents define the minimum installation, inspection, and testing requirements that the contractor must meet unless more stringent requirements are imposed by local, state or province, or national government. The more stringent of the two sets of requirements, which will usually be the project technical specifications, should establish the minimum installation, inspection, and testing requirements that must be met by the contractor.
- **Design and Installation Codes and Regulations.** Design and installation codes and regulations establish the minimum legal standard for installations. These codes and regulations are normally adopted by the local, state or province, or national government that has jurisdiction over the project being built.

    When applying design and installation codes and regulations, it is important that the correct edition of the particular code or regulation is used. In the United States, the latest edition of most codes is not in force until it is adopted by the governmental body that controls the project. In addition, many localities adopt codes with modifications, use only selected parts, or have a local code of their own, which is based on the national code. Contractors must make sure that documented installation, inspection, and testing procedures are in conformance with local codes and regulations.
- **Industry Standards and Recommended Practices.** Industry standards and recommended practices that specify installation, inspection, and testing procedures are normally not adopted by governmental bodies, but may be required to be adhered to by the project technical specifications. When industry standards and recommended practices are referenced in the technical specifications, contractors should obtain a copy of the document and adhere to any installation, inspection, or testing procedure specified.

Most developed countries have their own industry standards and recommended practices. When working internationally, contractors need to be aware of and adhere to the host country's industry standards and recommended practices. Knowledge of manufacturing standards and practices is also important when importing materials and equipment for use in domestic construction to ensure safety, compatibility, and performance. With the globalization of material and equipment markets, many countries and manufacturing organizations are adopting international standards.

- **Third-Party Listing and Labeling.** Third-party testing organizations test materials and equipment to ensure that it is safe for a particular application. An example is Underwriters Laboratories (UL). If the materials and equipment pass the stringent testing requirements, the materials and equipment are listed and labeled accordingly. This independent third-party testing has the advantage of eliminating the need for on-site inspection and testing of materials and equipment unless there is evidence of damage or modifications have been made.

  If approval of a material or equipment is based on third-party testing, then the contractor should be aware of the test methods and procedures to properly install the material or equipment.

- **Manufacturer's Instructions and Recommendations.** When installing materials and equipment, contractors should always install them in accordance with any instructions or recommendations provided by the manufacturer. Ignoring manufacturer instructions and recommendations may result in reduced performance and life expectancy of the material or equipment. In addition, manufacturer warranties and guarantees may be voided if instructions and recommendations are not followed.

- **Industry Custom and Practice.** Industry custom and practice varies throughout North America. Industry custom and practice is usually defined by accepted design and installation methods for similar types of construction in a particular area. In addition, sophisticated owners who the contractor works for may also have specific requirements that need to be addressed. When working in a particular area, contractors must be familiar with local customs and practices regarding the installation of materials and equipment.

How Should Production Procedures Be Documented?

Procedures should be documented as simply and as concisely as possible. The goal is to communicate the required procedures to those who have to perform them as effectively as possible. Written procedures are of course the standard method of communicating the required information. However, where more effective means are available they should be used. Flowcharts, diagrams, photographs, and other graphic methods should be used wherever possible to supplement or even replace written procedures. Videotapes illustrating installation, inspection, and testing procedures could also be used as well as computer simulation and animation if these mediums would more effectively communicate the procedure.

Using Documented Production Procedures

Documented installation, inspection, and testing procedures should be dynamic. One of the purposes of documenting these procedures is to provide a baseline for continuous improvement of construction processes. As new and better ways of performing the work are identified and adopted, the documented procedures should be reviewed and updated to reflect the current state of the art in construction.

FORMS, TAGS, CHECKLISTS, AND RECORDS

The purpose of forms, tags, checklists, and records is to facilitate the documentation of installation, inspection, and testing procedures in the field. Documentation of the completion and results of construction processes is a necessary part of any QA program. Documentation is required for compliance with the ANSI/ASQ Q9000 series of quality management and assurance standards.

Standard forms, tags, checklists, and records assist the construction firm in meeting the documentation requirements of its QA program. These standard documents save time in the field and help project personnel provide the data and information about construction operations in a consistent format. Properly maintained and filed, this information is then available for use internally or for meeting the customer's reporting requirements. Internally, these data and information can be valuable when used to solve project problems or as a feedback mechanism for continuous improvement of the firm's construction operations.

EVALUATING YOUR QUALITY ASSURANCE PROGRAM

You should evaluate your QA program to:

- Determine the conformity of your firm's QA program with specified customer requirements
- Determine the effectiveness of your firm's QA program in meeting your firm's quality objectives
- Provide the opportunity for feedback to improve your firm's QA program

Evaluation of your firm's QA program should be viewed as a positive learning experience regardless of whether the evaluation is completed internally, by a customer, or by an outside third party as part of a registration process. If nothing else, QA program evaluations keep QA at the forefront of the construction firm's consciousness.

Internal QA program evaluation should take place at regularly scheduled intervals. These internal evaluations should take place more than once per year. If possible, some type of internal QA program evaluation should take place quarterly. The frequency of internal QA program evaluation should be determined during the program development process and adhered to religiously. When possible, internal QA program evaluations should be planned during traditionally slow times when employees have more time to devote to the evaluation process.

Evaluation Criteria

Comparison of the documented QA program with what is actually being accomplished within the firm is the first step in self evaluation. If it is not in accordance with documented procedures, the reason for noncompliance must be determined. Once the reason for noncompliance is determined, either the documented procedure must be changed to agree with what is actually being done or what is actually being done must be changed to comply with the procedure.

The ability to meet customer expectations regarding the firm's QA program is also an important evaluation criterion. Feedback from customer assessments of the contractor's QA program will tell the contractor if the program meets the customer's requirements or if it needs to be modified.

The purpose of the QA program is to satisfy the customer by providing improved quality at a lower cost through increased productivity. The most important test of the effectiveness of the QA program is customer satisfaction. Customer satisfaction can be measured through customer feedback in a variety of ways that include written questionnaires and personal interviews. However, the most important feedback that the contractor can get is repeat business.

How Should Your Quality Assurance Program Be Evaluated?

ANSI/ISO/ASQ Standard 10011 provides guidelines for auditing QA programs. This standard is divided into three parts. Part 1 of the standard program deals with the auditing process. The qualification and selection of auditors is discussed in Part 2. Lastly, Part 3 addresses the management of the audit program. If the construction firm is interested in a structured approach to self evaluation, then it should obtain a copy of this standard and use it as the blueprint for performing self evaluations.

In general, the steps required to perform a comprehensive self evaluation include the following:

- Step 1—Define the scope of the evaluation
- Step 2—Identify the evaluation team
- Step 3—Develop an evaluation plan
- Step 4—Perform the evaluation
- Step 5—Prepare and submit an evaluation report

Internal personnel can make effective evaluators of the construction firm's QA program. If possible, however, the individuals reviewing a given department's QA procedures should be from another department to increase objectivity.

Outside consultants can also be used to evaluate the contractor's QA program. The use of outside consultants can be particularly valuable if the firm is seeking ISO 9000 registration.

Another construction firm from outside the contractor's geographic market area or area of expertise can also be an effective evaluator. This is especially true if the two firms take turns evaluating each other's QA programs. This peer evaluation should benefit both firms by providing feedback to the firm being evaluated as well as a look at how the other firm's QA program operates.

The evaluation team could be made up of a combination of these individuals. This approach takes advantage of the benefits associated with each of the various groups that can be used to perform internal evaluations of your QA program.

Should Your Firm Become ISO 9000 Registered?

The decision to pursue ISO 9000 registration should be made by comparing the costs and benefits of registration. ISO 9000 registration is time consuming and costly. Additionally, the investment in ISO 9000 registration is not a one-time cost, but requires an ongoing investment. In addition to the initial registration audit, there are ongoing third-party audits that must be performed on a regular basis to maintain registration. The advantages of ISO 9000 registration include:

- If your firm is currently doing business internationally or plans to expand its markets overseas in the near future, customers may require ISO 9000 registration. In short, ISO 9000 registration may be the passport your firm needs to enter the global marketplace.
- If your firm is doing business domestically with quality conscious customers, ISO 9000 registration may give your firm a competitive advantage over other equally qualified contracting firms. This is espe-

cially true if you are working for international customers who are also being required to be ISO 9000 registered.

- ISO 9000 registration may reduce the number of on-site audits and information your firm is required to provide quality-conscious customers. Depending on the extent and frequency of customer audits, eliminating these audits may offset the initial and ongoing costs associated with ISO 9000 registration.
- Preparing for and successfully completing the ISO 9000 registration process should result in improved performance throughout your organization.
- Knowing that once a quarter your firm's QA program will be audited by a third party to maintain your registration will tend to keep QA at the forefront of your employees' consciousness.

Getting your firm ISO 9000 registered can be viewed as a two-phase process. The first phase is to develop and implement an effective QA program. This can be accomplished by following the steps described in this chapter. Once the first phase of the registration process is complete, the construction firm can proceed with the formal registration process. The registration process can be summarized by the following five steps:

- Step 1—Select an assessor
- Step 2—Submit documents for audit
- Step 3—Undergo on-site assessment
- Step 4—Address any deficiencies identified
- Step 5—Receive registration

A current list of assessors in the United States can be obtained from the ANSI-ASQ National Accreditation Board (www.anab.org).

Registration can take anywhere from three to eighteen months or longer to obtain, depending on how dedicated the construction firm is to becoming ISO 9000 registered. Typically, the construction firm can expect to obtain registration in about six to nine months with concerted effort.

CONCLUSIONS

This chapter presents specific procedures that can easily be used by contractors to prepare effective QA programs. These guidelines were developed after identifying the needs and requirements of contractors for a QA program, and with the objective of providing a tool that can be an integral part of a construction firm's TQM program.

WEB ADDED VALUE

Q9001ManualSample.doc
Q9002ManualSample.doc
Sample Documents Folder

These files provide materials to assist contractors in preparing a QA manual and implementing an effective QA program.

This book has free material available for download from the
Web Added Value™ resource center at *www.jrosspub.com*

10

PARTNERING

Dr. James E. Rowings, *Kiewit Corporation*
Dr. Mark O. Federle, *Marquette University*

INTRODUCTION

As contractors examine ways to continuously improve their operations and stay competitive, the concepts of productivity improvement, quality control, and total quality management (TQM) provide internal processes for fine-tuning their organizations. The next level or evolution beyond these internal steps is to look to external processes, suppliers, and customers to gain a new level of bundled competitive advantage.

The concept of partnering is widely accepted within the construction industry as not only viable, but a fundamental project delivery system. General contractors working on public works, primarily under the U.S. Army Corps of Engineers, pioneered the formal concept in an attempt to reduce the amount of litigation being experienced on these projects. Basic concepts underlying partnering have come to the U.S. construction industry from Japan where it developed naturally as a product of the Japanese culture (their concept of competition is different from ours).

Partnering has been successfully adapted to both private and public sectors in the U.S. construction industry in the relationship between owners and contractors, contractors and suppliers, and contractors and subcontractors. Large private owners have used partnering to develop ongoing relationships that extend well beyond a single project and encompass an entire business unit for design and construction work.

Partnering represents the next frontier, the key to adding value for the customer. This chapter illustrates several ways in which partnering is being used in the construction industry and introduces a partnering model that fits varied needs of small contractors.

PARTNERING FUNDAMENTALS

The concept of partnering incorporates such fundamental ideas as long-term vision, respect, trust, and focus on common or at least compatible objectives among partners. The outcomes of long-term partnering are normally directed at achieving workload stability and reduced overheads. The approach is normally a maturing element of a successful TQM approach to work process improvement. In this way, Chapter 10 builds on our earlier chapters (Chapter 8, Total Quality Management and Chapter 9, Quality Assurance).

Formal research literature on partnering is somewhat limited in scope. Few articles describe any theory behind partnering since it is an application of several key management concepts. Many articles document success stories that typically involve large industrial owners and large industrial contractors. They validate the concept with documented results and a general explanation of the processes followed. Little, however, has been documented regarding the perspective of the small contractor.

Partnering is a team approach to project execution. In fact, partnering is teamwork at its best. You may notice that some of the concepts discussed in this chapter are not necessarily new. Many people may state that there was a time when these principles were part of the construction industry. We believe that as the speed, complexity, and competitiveness of the industry have increased, these concepts are not practiced on a routine basis. However, through the recognition and creation of a formal partnering process, any contractor can make good practice permanent and thus have a greater likelihood of sustained growth and performance.

The concept of partnering embraces the elements of shared vision, shared risk-taking, and mutual respect. These elements can only be achieved if certain principles are closely followed and attitudes are maintained throughout the project. The day-to-day challenges in construction and business environments constantly test for these elements. A real commitment and effort are required to keep these elements active so that the benefits of partnering can be realized, particularly because the team partners work together for a relatively short period of time. Each party must recognize the mutual benefit that long-term relationships—and with it long-term survival—can provide to all partners.

Partnering Background

The original use of the term *partnering* and the start of partnering concepts can be traced to the attempts made in the manufacturing sector with labor in the 1950s. The concept was to share risk and profit between the stockholder and the factory workers. Such approaches continue to be used in various forms through employee ownership programs and profit-sharing schemes.

The first modern-day partnering concept grew out of the Business Roundtable's Construction Industry Cost Effectiveness (Construction Industry Cost Effectiveness 1983). This effort was a joint study by both construction contractors and large industrial owners. The study recommended the formation of strategic alliances for design and construction of particular types of process plants. These alliances became common in the private industrial sector and today are often referred to as partnering as well as alliances.

The first use of partnering in the public sector was on a U.S. Army Corps of Engineers' project in the northwest region of the United States in 1989 (Fails Management Institute 1994). It occurred after the contract had been signed. The relationship was formed at an individual level rather than a corporate and agency level and used team-building principles. The idea was to address the areas where disputes might typically arise. This public sector model has been used many times by the U.S. Army Corps of Engineers and by many of the state transportation agencies since the early 1990s. Moreover, the private sector has seen benefits achieved in the public sector and has also adopted the basic concepts. Typically, more flexibility exists in the private than public sector. Although each of these developments has followed similar principles, various forms and timing are used for partnering.

Partnering today typically finds its way into the project development process either through a requirement established in contract documents or through a suggestion and voluntary commitment by the general contractor. The former approach may bring less than committed and reluctant participants into a process where there is no guarantee of a successful effort to establish a unified approach to the project. Several state transportation agencies follow this approach with mixed results. The latter approach, where the participants voluntarily participate, brings committed, or at least curious, participants to the process. This difference in attitude and orientation is crucial to the chances of success in obtaining understanding and alignment of the parties' objectives.

While partnering has evolved and exhibits various types of application, each partnering process examined has been built on the same basic underlying principles.

Partnering Principles

There are several common principles that the partnering process attempts to build. One goal of the process is to develop a set of common or at least compatible objectives that define the purpose of the team. This goal is the key element that bonds any team together. Sharing these objectives assures that each individual recognizes the outcome that is desired and that the group will not focus on outcomes that are at cross-purposes.

The idea in partnering is to develop a set of objectives that allow all stakeholders in the process to achieve their objectives without doing it at the expense of another. Without common objectives, a win-lose relationship will likely flourish, as competition is developed to maximize one's own objectives. For a partnering process to work, it is vital that subcontractors, contractors, suppliers, designers, and clients do not develop positions where one wins at another's loss.

The win-lose situation typically occurs when claims are involved, as each side attempts to bias their request to establish some movement for negotiation purposes. Rather, a common set of objectives for all must be set. The process begins by identifying what each party would like to achieve and at the same time defining items that would represent a loss. By communicating these objectives, parties have an opportunity to discuss and clarify them. The partnering process should allow each participant to identify individual and corporate objectives and develop common language to describe collective objectives for the team.

A second guiding principle is that the group should work toward team success. True partnering builds a commitment to team success. Goals are aligned so that success is dependent on each party achieving its goals. Thus, each party is committed to doing everything within its power to help others achieve their goals. The project will not be successful if only one part of the team gains advantages or is profitable; all must reach their goals and objectives—the owners, contractors, suppliers, and labor—in a total partnering sense.

This principle rests on first having a set of common goals for all project participants. Where organizations have worked together for many years, they typically begin the process of partnering with the establishment of common objectives and commit to team success through the process. The process for many construction projects may be like this in the private sector where there has been a consistent, positive ongoing relationship between the parties. However, in many projects, the parties are unfamiliar with each other and must establish the foundation before the first two principles of partnering can be realized. Steps to establish these foundations must sometimes be a part of the partnering process.

Trust

The pivotal ingredient to make partnering work is the issue of trust. Trust creates an environment where individuals are willing to openly share all required information. It allows people to provide both positive and negative news so that appropriate corrective action can begin to put the team back on a path toward common goals and objectives.

Trust begins with risk taking and builds through experience. For trust to spread requires that a leader start the process by demonstrating trust to subordinates. This is the pattern for others to follow and reciprocate. It typically starts at the top with the general contractors or the owner and spreads through the organization. In some cases, where there are several authority relationships, trust must be developed in each. This is the case of regulatory agencies and private owners with contractors and labor union leadership and contractors with the workforce. Trust is founded in confidence in the ability of another to do the right thing, be fair, and perform to desired standards. Trust is typically granted in a stepwise fashion and is constantly monitored and evaluated. It grows as long as expectations are realized. Where expectations are exceeded, greater trust is awarded. Where expectations are not fully realized, trust is withdrawn.

Trust is arrived at through communication early in a project followed by repeated high performance. Getting expectations clearly established and communicated to all parties is the key to establishing the trust relationship.

Risk Sharing

Another foundation block in developing partnering is risk sharing. Sharing risk, rather than avoiding risk or "sticking" the other guy, forms the basis of a relationship. If participants feel they are being set up for failure or are bearing risks that they cannot estimate or manage, they will be unwilling to compromise or assume additional responsibilities that could benefit other members of the team. This put them in a survival mode where they will start to look at every situation as a win-lose opportunity. The behavior will be selfish rather than team oriented. Specialty contractors have found themselves in this position with indemnification clauses imposed on them by owners or general contractors. Market conditions in the competitive bid sector often create this type of situation.

The intent of some of the partnering processes is to bring these issues to the surface and find ways to share risks or reassign them in a more equitable manner. Partners agree to distribute and in some cases share risks where more than one party controls the risk. Risks should be appropriately compensated for, within current market conditions, for those willing to assume these risks. Communication processes give parties the ability to understand the nature of the risks and provide alternative ways that the risks can be managed or shared.

Risk sharing begins with an understanding of each other's needs. Such understanding can only occur in an environment where individuals are committed to having a relationship for the duration of the project. Communication of one's feelings, needs, and goals typically occurs after a personal relationship has been established. The personal risk and investment required to communicate such items usually dictate that the individuals share a common ground of understanding at the personal level. Proximity and repetition can both contribute to building this type of bond so that one can understand the other's feelings, needs, and wants.

The process should begin to establish what individual team member's needs are and what corporate needs they are representing within the team. Partners share their own challenges and the other person's potential barriers to success.

As important as identifying needs is to discuss the fears or barriers that individuals face. Through identifying these barriers, the team can seek ways to avoid pitfalls and put systems and processes in place to achieve the objectives. Partners work to develop a plan that can lead to success for both.

BENEFITS OF PARTNERING

Benefits of the partnering approach include more efficient relationships and transactions that occur in a team environment. Eliminating nonvalue-added activities associated with conflicts and conflict resolution represents a significant cost savings. Energy can then be directed toward value-adding activities. These activities are the major driving force for entering into partnering. Improved relationships and understanding use less time for clear communication. Moreover, they reduce the amount of time spent on posturing for potential negotiation through some form of dispute resolution.

Improved relationships also create teams with less reliance on inspection and rework for acceptance. This concept is one that the entire TQM process is built around. Costs for repeated inspection add no value, but only consume financial resources. With partnering in place, the commitment is made to deliver a product that can be processed through the next step toward the final product without first needing to stop and inspect it.

Partnering also puts fun back in work through better working relationships and job satisfaction from group accomplishment. People want to enjoy their work. The type of working relationships they have affects how they feel, how much stress they have, and how they encourage others to follow their footsteps. The construction industry has had a history of conflicts, stress, and negative attitudes toward careers in construction. Partnering can reverse this trend and provide an improved climate for recruiting future talent.

PARTNERING APPLICATIONS

Several partnering applications exist in practice. By far, the most common application is the project partnering agreement where the general contractor and owner hold a one- to three-day meeting to build the on-site team and develop a partnering charter, which summarizes the agreements made during the partnering process. These agreements take many forms. They depend on assumptions made concerning the point that participants are relative to the establishment of a foundation for partnering; the level of trust and expected risk sharing; established personal relationships; and the understanding of individual feelings, needs, and goals.

For individuals who have generally worked together previously, a one-day workshop may be sufficient. Longer sessions are used when it is necessary to build a team from individuals who have never worked together before. In these workshops, it is often useful to use some form of standard personality assessment tool to provide a way of getting people to analyze themselves and others in a common form. This provides a way for individuals to talk about themselves and learn about others in a more personal way. There are several common tools used for this purpose with varying costs, expertise, and time requirements to administer (M. Liteman, Campbell, and J. Liteman 2006). Each was developed for a specific purpose, none of which was specifically for the partnering process. These tools provide insights in how people may react in certain conflict situations (Strength Development Inventory) and personality traits and tendencies (Myers-Briggs Type Indicator).

For shorter partnering approaches, the facilitator will use a combination icebreaker activity and introductions of participants to establish an entry into understanding an individual's feelings, needs, and goals. For example, a facilitator may ask individuals to interview each other and provide an introduction to the group that includes something of a fun or personal nature about the person (i.e., favorite sport or hobby, something unusual about the person, or any type of animal that would be descriptive of the individual). Obviously, one must have some sensitivity of the feelings of the group when using these methods so as not to embarrass or insult anyone. Humor can be a benefit but also carries some risk and should be used sparingly in these situations.

The goal of each session is to begin the team-building process in an accelerated way. The objective is to reach several mutual commitments related to the project. The charter is the document that captures the understanding developed in the workshop. Each charter takes a different form. If too strong of a charter example is given to the group, the individuals will not be creative and develop their own unique charter. Participants should carry a commitment to the charter. It is best that the charter represents the group's language. The charter will also reflect the leadership and strength of opinions within the team in the selection of the words, phrases, and priorities. These workshops are a logical next step to the TQM efforts that many forms are engaged in. Much of the language in the charter and in the discussions can be directly tied to the TQM process and methods.

The general contractor is typically in the lead for project partnering. We believe that it would be difficult to have partnering on a project without the cooperation of the general contractor. However, there might be instances where the partnering concept could be brought to the general contractor for consideration by a specialty contractor. In cases where multiple prime contractors exist, it would be advantageous to suggest a partnering approach to the owner directly. Partnering can always be applied with the specialty contractor's subcontractors and suppliers for a project, regardless of the general contractor's view.

The evolution process that partnering workshops sets in motion is depicted in Figure 10.1. There are three distinct stages in the evolutionary process. The initial stage is when there are independent entities. This can be described as the point when goals are independently driven; each entity has its own self-sustaining systems; and the relationship and transaction is short term and nonexclusive. The relationship is typically formed in a competitive environment and may even be adversarial in nature. The processes are designed to be independent, with inspection and verification of performance between each of the points of exchange. Goals may actually conflict, such as the case where the owner's minimum standards of performance are also the maximum standards of performance for the contractor.

The second stage of the evolution is when there are aligned expectations. This stage is typically the goal of the partnering workshop. This stage is exemplified by having risks and levels of trust clearly established. At this stage, parties will have complementary goals and compatible systems to manage the work processes. The value set will be shared by the parties and customized to project requirements. The organization will still typically be reactive to problems rather than proactive to opportunities. The planning cycle will be relatively short term and not compatible beyond the individual's time boundaries for the project.

The last stage in evolution is where the work processes are fully integrated. In this stage, the entities exhibit strong teaming behavior, share risk, and have fully

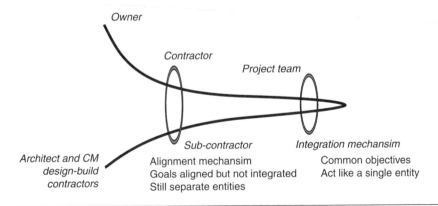

Figure 10.1 Partnering evolution

integrated management. Systems for work processes appear externally to have transparent interfaces. Success is identified with the combination rather than the individual component organizations. The entities seek opportunities together and actively market bundled services and products.

The alignment mechanism is the partnering workshop and resulting charter. The integration mechanism only occurs when individual parties share financial risk or establish an alliance that extends to continuous future working relationships. Typically the project partnering process can achieve only the second stage of evolution. For the specialty contractor, there are several opportunities for exceeding this level and reaching the final stage. These include contractor-owner partnering, contractor-supplier partnering, and contractor-labor partnering.

Several other applications of the partnering concept are possible. These include the development toward contractor-supplier partnering relationships. We have found many examples of partnering concepts being applied, but none where a formal partnering process had been used to establish an exclusive long-term, stage-two type of relationship. Many contractors have strong relationships with certain suppliers. In some cases, they have electronic linking between offices for checking inventory, pricing, and ordering materials. Others have developed special relationships between their work processes to reduce packaging and handling costs associated with supply. In other cases, they have addressed the process of getting paid and dealing with returns in exchange for preferred pricing and treatment. In each of these instances, common objectives, shared risk, and trust were key to bringing about savings to each party. Opportunities evolved slowly over time. While not developed through a formal process, the workshop concept could be applied with a group of key suppliers to develop a set of supplier-contractor relationships with features similar to those identified earlier.

Other opportunities for integrating work processes and systems could produce significant competitive advantage. The one area and work process that might provide the greatest opportunity for partnering is the marketing effort. This partnering concept can even be taken to the level of dealing with the manufacturer as well as the supplier. The opportunity to take this relationship to the third stage of the evolution process may exist as the producer and installer bring their goals into alignment.

A second area for specialty contractors to apply partnering concepts is with the relationship that some firms have with owners in their service divisions. Some examples could be seen where there were trust and common objectives between parties developed over time. This type of work lends itself to the development of long-term relationships involving suppliers, contractors, and owners. The workshop model could be used to develop opportunities and identify areas where risk could be shared.

The labor-contractor area is one with tremendous potential. This is by far the most challenging yet potentially rewarding area for specialty contractors. The traditional relationship between the contractor and labor is an adversarial one in which they are each trying to get a greater share of the common pie. This relationship orientation is clearly stage one in the evolution process.

Mission Statement

To start aligning common goals a group needs a mission statement that reflects their belief regarding the task that brought them together as shown in Figure 10.2. Many times, the group mission statement results in a mission statement for the project. An example of a mission statement that might result would be the following:

We are a team dedicated to providing a quality process in accordance with our contract. We are committed to safe, economical, fast, high quality work processes.

Further information regarding mission statements can be found in Chapter 8 on Total Quality Management.

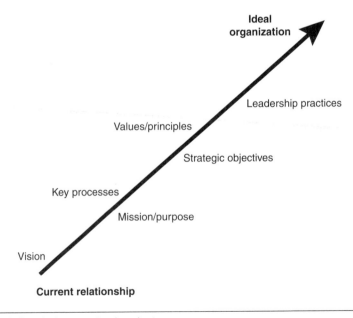

Figure 10.2 Developing a partnering charter

Examples of outcome objectives at an early stage of partnering might be as follows:

- Complete the project without disputes
- Complete projects at a cost that provides reasonable profits for all parties, fair treatment of all employees, a positive working environment, and a reasonable price for the completed facility
- Complete the project without any external intervention in disputes

Examples of process objectives for an early stage of partnering are as follows:

- Make decisions at the lowest level possible
- Hold pre-job conferences to identify and resolve jurisdictional issues
- Maintain open and honest communication

Facilitator Use

The use of an outside facilitator is extremely important to making the partnering workshops productive and positive. The facilitator should fit the culture of the company and should be selected through an interview process rather than solely on the basis of price. He or she should be familiar with the construction industry and with any special circumstances within the team. If contracts are already in place, he or she should be familiar with the roles and responsibilities. The facilitator's role is to help participants to share expectations, needs, feelings, knowledge, and attitudes with others. He or she helps people communicate clearly and makes sure feedback is provided for everything developed for the charter.

Facilitators should seek out areas that have not been discussed but should be. They must read body language and understand when to seek comment and draw out opposition where it exists. The facilitator normally identifies the ground rules for communication in the workshop and serves as a referee for these rules. He or she must protect individuals from personal attacks when differences exist, guide the group toward the objective, and record the outcome.

Finally, the facilitator must organize and manage the use of time. Using ground rules and being results-driven, they push the agreed-upon objective for the meeting. The more experience that a facilitator has with workshops, the better he or she can develop specific techniques to handle situations that arise.

In many cases, the size of the group may dictate that a team of facilitators and support staff is used to manage the partnering process. Facilitators must appear to all parties not to favor, be subordinate to, or be biased against any participant. It is hard to have this objectivity when there is a working relationship with part of the team involved in the partnering process.

The many sources for facilitators include local faculty from universities and community colleges, management consulting firms, and counseling services. A facilitator checklist has been developed by the Associated General Contractors (AGC) of America (McIntyre 1995).

RECOMMENDED STEPS FOR IMPLEMENTATION

The partnering process has been documented in many trade journals and other publications. These articles have explained the benefits and described the positive outcomes that resulted from the process. Contractors should look to this process as a consistent and disciplined way of communicating needs and expectations and gaining the same from other members of the team. The process can be applied in a number of situations including:

- Owner-contractor
- Contractor-contractor
- Contractor-supplier
- Management-labor

Each of these will involve different parties with different needs for the specific time and situation. The process with the workshop format can be customized and used in each situation. The first step is using partnering to commit to the concepts and select a place to begin applying it to company operations.

We recommend the process be tried first in an internal situation. This provides an immediate benefit and allows the contractor to select and evaluate an initial facilitator. It allows the facilitator to get to understand more about the company before tackling external situations. Internal partnering can be applied to a major project that is coming up to improve communication links between the field and the office and make sure the contractor has a unified team going into the project.

The workshop can be used to address improvements of one particular aspect or work process. It can be used to address roles and responsibilities within a new organizational or ownership situation. Each of these items provides an opportunity to strengthen the company's internal team and achieve an improvement in competitiveness, efficiency, and cost-effectiveness. Internal partnering provides a forum for people to encourage communication where there may be reluctance to discuss some issues with all parties involved. Normal authority relationships within the company must be relinquished to the facilitator if open communication and sharing is to occur.

Owner-Contractor

Individuals with the appropriate level of responsibility in the owner's organization should be approached with the concept. In preparing for the presentation, a list of potential benefits to the owner should be compiled. Present a plan on how the two firms could conduct the partnering process and offer to make arrangements and organize the workshop. If there is a favorable response, establish a time for a planning meeting or conference call with the facilitator. This meeting is important to establish the objective for the workshop and the necessary participants and parties to be represented. It also provides an opportunity for the facilitator to understand more about the partnering environment.

Each party to this meeting will leave with certain responsibilities. Representatives for the owner and the contractor will be responsible for communicating expectations for the workshop to key participants in their organizations and for making the appropriate detailed agenda and plan for the workshop. One of the individuals will be responsible for making the meeting arrangements and other logistical decisions pertaining to the workshop.

Contractor-Contractor

Another opportunity for partnering is with other trade contractors or with the general contractor. These situations can be for a specific project or for a long-term relationship. As shown in Figure 10.1, the first step is to develop the partnering relationship on a single project, then extend the process to develop more interdependence. The workshop is used to communicate expectations and establish common objectives and strategies. The organization process is similar to working with the owner. The greatest benefit of these workshops can occur before there is a contract between the parties. Contract language often constrains opportunities available to workshop participants. The selection process, if competitive, can also create an adversarial relationship that is not conducive to partnering.

Contractor-Supplier

The next logical area for contractors to approach with partnering is with primary suppliers. These may be material suppliers, equipment suppliers, or service providers, including designers, accountants, repair shops, and others. The usual motivation for the supplier is to gain an exclusive relationship for providing the service. The challenge in getting the individual to listen to a proposal to partner is much less since the contractor is the customer. Establishing the partnering process is the same as with the contractor-owner. Attention should still be given to developing a list of benefits for the suppliers prior to approaching them with the concept.

Management-Labor

The last area where implementation is possible is with labor. This can be approached on a single project or a long-term local area arrangement. For a single project the process only needs to apply for a specific period of time and with limited involvement. This type of partnering is much like that between contractors. The long-term local union area form of partnering is far more complex due to the dynamic nature of the market conditions governing supply and demand for labor, the potential for changing leadership within the unions, and the implications to all members within a collective bargaining unit for management.

Expectations for partnering between management and labor should not be overstated. Obstacles are the very foundation elements of partnering, like trust, risk sharing, and common objectives. Adversarial relationships established at the bargaining table over the years are real impediments to getting started. On both sides leadership is required and must sense a potentially greater benefit than the risk. Threat of continued erosion of union construction in an area or the desire to expand the union construction market is an area of common ground for entering into the process.

Leaders need to discuss and agree on reasonable expectations to start the process. They also need to identify the appropriate type of participants—those who have a heavy stake in the outcome, are open to listening, and are recognized as leaders by others. They are not representing others, but are representing themselves. However, they will set a pattern that can be voluntarily followed or made common practice in the future. Members of the bargaining committees are not the appropriate participants to start with, although some members would be acceptable. The business manager is a must, but that person should be receptive to the concept. Since the business managers are elected, they obviously have some political strength within the local unions. Timing is everything with this form of partnering. The business manager who is soon up for election will not share risk. It is much easier to share risk and trust the other side between contract negotiations and away from elections. Also, market conditions in the area will affect the attitude toward partnering. It is best to get the well-respected veterans in each organization. Assistance from regional or international representatives for the unions is also beneficial. If there is encouragement at this level, pressure on local officers and business managers can be relieved somewhat.

Management representatives also need to be carefully selected. Individuals should be well respected by their fellow contractors, but also respected by the unions, if possible. Individuals who can best relate to union representatives are those with experience as craftsmen or with significant time working in the field.

The individuals should not have any current grievances pending or be actively involved in some form of negotiation or adversarial action with other members of the management group.

The process for management-labor partnering should begin with a planning meeting for the workshop. Our experience indicates that two workshops within several weeks of each other is the best approach. The first workshop serves to create an awareness of partnering, establish certain ground rules, and establish several directions for future workshop sessions. The time between the first and second workshops allows initial participants to persuade other key participants to attend the next session. Also, several weeks are close enough that the group does not have to revisit all of the first discussion before proceeding. The second workshop should have the goal of developing a draft of vision and mission statements for the partnering offer. If time and willingness exists, the group can also discuss strategies that can be used. The second session should end with measurable commitments and actions for both sides to take away from the workshop. For significant results, the process will require an ongoing and broader involvement from both management and labor.

CONCLUSIONS

Partnering offers contractors a significant opportunity. As a process, it can be used to align objectives for a single project or align two or more parties in a long-term relationship. The value of the facilitated workshop is to start the process and provide a disciplined, formal time and place for the exchange of information about needs, feelings, expectations, and knowledge. The goal of the workshop is to reach agreement on objectives and strategy. The partnering process is intended to produce common objectives, trust, and risk sharing.

Conversations with contractors have revealed that there are owners who use partnering to get a commitment from contractors not to submit claims, no matter how unreasonable the owners are in recognizing their obligations under the contract. Similarly, there are owners who referred to being "partnered." They feel some contractors use partnering to dodge necessary contract requirements. Risks with partnering are almost wholly associated with the issue of trust. Therefore, the key to the effectiveness of the partnering process is not the workshop or the charter. It is rather the daily commitment and activity that supports the team's established goals.

WEB ADDED VALUE

Partnering.pptx

This file provides a template for a partnering workshop. The template includes a sample partnering agenda, an introduction to the basic partnering concepts, sample mission statements, and a sample partnering charter. This template is intended to provide a starting point for a contracting firm to develop its own documents to implement the workshop process presented in this chapter.

11

PERFORMANCE EVALUATIONS

Dr. Awad S. Hanna, *University of Wisconsin-Madison*
Jacqueline M. K. Brusoe, *University of Wisconsin-Madison*

INTRODUCTION

Because of competition, contractors must continually search for ways to improve performance and productivity. One way to improve company performance is by enhancing the job performance of individual workers, whether supervisors or craftsmen. Employees can improve when they have a clear understanding of how well they are working and how this compares with a standard of quality performance. One of the best ways to give workers an understanding of their work quality is through job performance evaluations.

The goal of this chapter is to assist contracting companies in the development of their own performance evaluation systems. Employee evaluations are widely used in all industries as a measure of performance. Information provided by performance evaluations can have significance for the construction industry, in particular for specialty contractors. Notably, in an industry as highly competitive and labor intensive as construction, employee performance can have a strong impact on a company's performance. Tracking employee performance through regular evaluations can (1) help monitor skill and productivity improvement, (2) pinpoint training needs, (3) determine proper skill assignments, and (4) deal with other aspects that affect employee performance.

Use of performance evaluations can result in both corporate and personal benefits: the company has a means by which to measure employee performance and the employees have defined standards and expectations to strive toward. One contractor improved his company's productivity 20 percent in the first year of implementing an evaluation program. He also claimed that establishing an evaluation form had improved morale, provided incentive and motivation for company workers, and helped the company's growth and reputation.

METHODOLOGY

The scope of this study was the investigation of performance evaluations in the electrical construction industry. This chapter recognizes the characteristics of performance evaluations in this industry and identifies those criteria (based on job duties and worker attributes) that can be used to evaluate the job performance of construction supervisors and craftsmen. This study did not intend to develop any sort of model which would demonstrate that certain, specific evaluation criteria would produce a desired job performance. The results presented in this chapter are a culmination of information offered by industry personnel. The organizations included in this study belong to the National Electrical Contractors Association (NECA). The participating companies were chosen randomly, regardless of their size, regional location, and type of work performed. To produce results that are representative of the industry, many types of contractors were surveyed for information.

The main objectives of the study were to (1) determine what type of construction company conducts performance evaluations, (2) ascertain details about the evaluations conducted (i.e., how frequently they are conducted, how long they take), (3) determine the typical job duties of supervisors and craftsmen, (4) determine what types of attributes are important in the job performance evaluations of construction supervisors and craftsmen, and (5) survey opinions regarding the reasons for which performance evaluation results should be used. To address these objectives the study began with an extensive investigation of literature on topics relating to job performance evaluations in the construction industry. While this was being completed, several interviews were conducted with personnel in local construction companies to establish preliminary criteria by which supervisors and craftsmen are evaluated. This information, along with that gathered in the literature search, was used to create two surveys. The purpose of these surveys was to solicit data regarding performance evaluations from electrical construction companies throughout the nation.

PURPOSE OF PERFORMANCE EVALUATIONS

For the past 20 years, performance evaluations have been accepted in many industries as a standard practice for assessing individual work effort. Experts estimate that approximately three-fourths of U.S. companies employ performance evaluation programs. The traditional uses of performance evaluations include determining the training needs of the staff, providing a basis for personnel actions, and motivating employees by providing feedback (Moss 1989). They can also serve the purpose of prodding supervisors to keep a closer eye on their subordinates, getting supervisors to hone their "coaching" skills, or improving the organization by identifying employees who possess promotion potential or require further development (Oberg 1972).

To gain maximum benefit from a performance evaluation program, managers must develop a quality program. Quality employee job performance evaluation programs should be designed with the purposes of providing (McGregor 1987):

1. Documentation to back up salary increases, promotions, transfers, and demotions
2. A means of telling employees how they are doing (suggesting needed changes in employees' behavior, attitudes, skills, or job knowledge; and letting the employees know "where they stand" with the boss)
3. A basis for the coaching and counseling of an individual by his or her supervisor

The first listed purpose benefits the organization, and the other two benefit the individual. Employee appraisal programs usually help improve the performance of both parties.

Performance evaluations can be used as a problem-solving tool to gain information about the company's organizational processes. Evaluations can also facilitate discussions between supervisors and employees about the nature of their jobs and factors influencing their job performance.

An evaluation program should be designated in accordance with the following four principles (Moravec 1983):

1. Substantiation—Supervisor needs to collect evidence to demonstrate the employee's output and overall performance since the previous evaluation before the evaluation session is held with the employee.
2. Development of understanding—Supervisor needs to clarify what is expected from the employee, and the employee needs to understand the supervisor's expectations. This mutual understanding allows the supervisor to compare the employee's work to expectations and make suggestions for future job performance.

3. Acknowledgment—Employee and supervisor need to agree that they have discussed relevant issues, regardless of their agreement on the issues themselves.
4. Action planning—Employee and supervisor define steps together to improve the employee's future job performance.

From a corporate standpoint, feedback from the evaluation can help create an employee who is more committed, motivated, and productive, a benefit that will ultimately help the company. On a personal level, the evaluation can help the employee achieve a renewed career direction, higher job satisfaction, targeting skill development, and greater career advancement. Productive, motivated, and satisfied workers who are continuously improving their skills can increase company profits and reputation.

Supervisors need to be involved in an employee's job before they can make an adequate determination as to whether or not the employee has fulfilled his or her job requirements. Research confirms the importance of this need. A study by Maloney and McFillen (1987) looked at which of a foreman's performance factors had the greatest influence on craftsmen. The study defined a foreman's job success as "his ability to motivate his employees to perform the required work." This study identified three factors that result in better job performance and more job satisfaction among craftsmen: (1) ability to assign crews whose abilities matched the job requirements; (2) willingness to participate with and support craftsmen; and (3) ability to properly facilitate the work to completion.

Another research study by Lemna, Borcherding, and Tucker (1986) has demonstrated that the following are traits of productive foremen: (1) sharing information with crews; (2) planning jobs in advance; and (3) regarding craftsmen preferences when assigning crews. In other words, supervisors who are familiar with their employees' everyday duties and are supportive can help produce better workers.

It is impossible to develop the performance of an individual without first specifying relevant "output units" to measure his or her performance. Developing suitable criteria and tolerance limits for evaluations may not be simple. While criteria for performance appraisals have usually been established through task assessment and structured job descriptions, managers should realized that these benchmarks may have shortcomings. Tasks and job descriptions can change over time, employees may develop other ways to do the job, and uncontrolled outside influences may affect the employee's job performance.

One way to realistically analyze the different aspects of an employee's job is to build a model, or a behavioral description, of an employee's ideal performance. This model must include four components (Robinson and Robinson 1996):

1. Performance results—Outcomes an employee must achieve to attain a goal. An employee's job may contain several performance results.

Each result needs to be labeled and defined to ensure an understanding of what is expected.

2. Best practices or competencies—Requirements of the employee if he or she is to successfully achieve the performance results. Best practices are what the best performers actually do in their job to achieve excellent results. Competencies describe the skills and knowledge required to produce satisfactory job results.

3. Quality criteria—Criteria used to measure the quality with which the job is performed need to be specified.

4. Work environment factors—Model of employee behavior needs to include forces inside and outside the organization that help or hinder the accomplishment of performance results.

For the proper evaluation of construction personnel, contractors should select performance measures that account for outside factors (working conditions, quality of equipment and materials, personality compatibility of crew members, realistic productivity goals, and so forth). This is especially true if an incentive system is used to reward outstanding job performance. Unless the performance exhibited by an individual or a crew can be defined and measured in terms of physical output, incentive programs based on money rewards will be ineffective.

SURVEY OF SUPERVISORS' DUTIES AND ATTRIBUTES

A survey was developed to gather information on supervisors. One thousand surveys were mailed to NECA contractors. One hundred and ninety-three were returned (19.3 percent response rate). Forty-seven percent of respondents held the position of president and 13 percent were vice presidents.

For the purposes of the survey and this chapter, the term *supervisor* is defined as the senior person on a construction site responsible for directing all foremen and craftsmen. There are many possible job titles for a person in this position—superintendent, supervisor, project manager, general foreman, among others. The title may depend on the size of the company and the project.

Supervisor Duties

The supervisor survey respondents were given a list of 29 potential supervisory duties. Each respondent was asked to place a check mark next to those duties he or she felt described the job of a supervisor in his or her company. Space was provided in which respondents could add job duties not included in the list. The duties that received above 80 percent response rate were based on two main factors: planning and communication. These duties are:

- Maintain working relationship with job's owner, general contractor, architect/engineer
- Determine crew requirements
- Review equipment and material needs for a job
- Incorporate safety into all work plans
- Participate in pre-construction meetings
- Review plans and specifications prior to each job
- Coordinate material needs with staff to ensure timely delivery
- Participate in developing project schedule

Those that received 60 percent and 70 percent response rates focus on the everyday duties associated with the upkeep of a job. These duties are:

- Keep track of equipment throughout project duration
- Review daily time sheets
- Conduct regular crew meetings
- Maintain up-to-date plans and specifications for the job
- Conduct regular safety meetings
- Maintain complete as-built drawings
- Review safety manual with crew prior to job
- Track, document, and report back charges for job
- Implement change orders
- Complete regular field reports

It is important to consider the duties a supervisor has when determining what type of criteria to use in his or her evaluation. These survey responses indicated that the duties listed are viewed by industry personnel as important aspects of a supervisor's job; these aspects should determine the criteria used to evaluate a supervisor's job performance.

Supervisor Attributes

The survey also asked respondents to rate the importance of certain attributes that may be used in the evaluation of a supervisor's job performance. Respondents were given a list including items in five categories: (1) management skills; (2) attitude toward work; (3) personal skills; (4) experience and/or training; and (5) job-related skills. The respondents were to circle the number they thought best represented the importance of each attribute in each category. The selection of ratings was as follows: 1 (extremely important), 2 (somewhat important), 3 (not too important), 4 (not important at all), and 5 (does not apply). Respondents were also asked to rank the five separate categories in numerical order, indicating which category was most important, second most important, and so on. The result from this survey section is a list of important attributes that should be considered when evaluating a supervisor's job performance.

Management Skills Attributes

The first category in which respondents were asked to rate attributes was the management skills category. This category's list was composed of the following; leadership, ability to instruct, ability to motivate, budget control, communication skills, delegation of responsibility, ability to negotiate, goal setting, public/customer relations, and ability to set a good example. The average ratings given to each attribute in this category were statistically analyzed to detect any significant differences. The attributes that stood out as being ranked the best were:

- Leadership ability
- Sets good example
- Ability to motivate
- Communication skills
- Public/customer relations
- Delegation of responsibility
- Ability to instruct

The ratings given to each attribute in this category are listed in Table 11.1.

Table 11.1 Management skills ratings

Criteria	Average rating
	(1 = Most important, 5 = Least important)
Leadership	1.11
Sets good example	1.19
Ability to motivate	1.23
Communication skills	1.34
Public/customer relations	1.37
Delegation of responsibility	1.42
Ability to instruct	1.43
Goal setting	1.73
Budget control	1.99
Ability to negotiate	2.00

Attitude Toward Work Attributes

In the attitude toward work category, respondents were asked to rate five attributes, which were ranked in the following order:

1. Work ethic
2. Attendance
3. Attitude
4. Personal conduct
5. Appearance

The ratings given to this list are displayed in Table 11.2.

Personal Skills Attributes

The attributes to be rated in the personal skills category were initiative, follows rules, dependability, organization, fairness, rapport with workers, accepts responsibility, creativity, trustworthiness, listening skills, and ability to work with others. A statistical analysis of the responses received by the 11 attributes showed that the following 5 were given the best ratings:

1. Trustworthiness
2. Dependability
3. Initiative
4. Accepts responsibility
5. Ability to work with others

Table 11.3 shows the ratings received by each attribute in this category.

Experience and/or Training Attributes

The following list was contained in the experience and/or training category: knowledge of work, productivity, technical capabilities, quality of work, versatility, ability to catch mistakes, and ability to communicate technical information.

Table 11.2 Attitude toward work ratings

Criteria	Average rating
	(1 = Most important, 5 = Least important)
Work ethic	1.06
Attendance	1.09
Attitude	1.09
Personal conduct	1.18
Appearance	1.55

Table 11.3 Personal skills ratings

Criteria	Average rating
	(1 = Most important, 5 = Least important)
Trustworthiness	1.09
Dependability	1.10
Initiative	1.15
Accepts responsibility	1.16
Ability to work with others	1.22
Listening skills	1.31
Follows rules	1.36
Organization	1.38
Fairness	1.41
Creativity	1.52
Rapport with workers	1.56

Table 11.4 Experience and/or training ratings

Criteria	Average rating
	(1 = Most important, 5 = Least important)
Knowledge of work	1.17
Productivity	1.27
Ability to catch mistakes	1.28
Quality of work	1.28
Ability to communicate technical information	1.39
Versatility	1.43
Technical capabilities	1.53

When a statistical analysis was performed on the responses to these seven attributes, four were viewed as most important:

1. Knowledge of work
2. Productivity
3. Ability to catch mistakes
4. Quality of work

Listed in Table 11.4 are the ratings given to each attribute in this category.

Job-Related Skills Attributes

Fourteen different attributes were listed in the job-related skills category. They were safety awareness, maintenance of records, planning, scheduling, cost awareness, change order management, site operations management, ability to deal with (job-related) problems, project closeout, participation in quality improvement process, communication with crews, project estimation, crew training, and ability to obtain/retain new work. Out of these 14 attributes, 6 were rated significantly higher than the other 8. These were:

1. Ability to deal with problems
2. Safety awareness
3. Planning
4. Communication with crews
5. Scheduling
6. Maintenance of records

Table 11.5 lists the ratings given to each attribute in this category.

Ranking of Individual Categories

After the respondents rated all of the attributes in the separate categories, they were asked to rank the categories themselves. According to the averages tallied from all of the responses, the order of importance (from most important to least important) of the categories was:

1. Management skills
2. Attitude toward work
3. Experience and/or training
4. Personal skills
5. Job-related skills

A statistical analysis showed there was a significant difference between the rankings of the first two categories and the last three, showing that respondents felt management skills and attitude toward work were significantly more important than experience and/or training, personal skills, or job-related skills. The ranking given to each category is shown in Table 11.6.

Goals of Performance Evaluations

The final section of the supervisors' survey asked why construction personnel think job performance evaluations should be conducted. Survey participants were given a list of 17 statements reflecting potential goals of performance evaluations. They were to give each statement a numerical rating indicating its importance as a goal for performance evaluations. The ratings from which respondents could select were 1 (extremely important), 2 (somewhat important), 3 (not too important), 4 (not important at all), and 5 (does not apply).

Table 11.5 Job-related skills ratings

Criteria	Average rating
	(1 = Most important, 5 = Least important)
Ability to deal with problems	1.22
Safety awareness	1.26
Planning	1.32
Communication with crews	1.39
Scheduling	1.45
Maintenance of records	1.48
Site operations management	1.58
Cost awareness	1.61
Project closeout	1.63
Change order management	1.78
Participation in quality improvement process	1.91
Ability to obtain/retain new work	2.06
Crew training	2.11
Project estimation	2.47

Table 11.6 Ranking of individual categories

Category	Average ranking
	1 = Most important, 5 = Least important
Management skills	2.08
Attitude toward work	2.16
Experience and/or training	2.78
Personal skills	2.92
Job-related skills	3.28

Analyzing these ratings gave the research team an idea of what industry personnel think job performance evaluations should accomplish. In reviewing the goals that received the best ratings from the survey participants, the research team found that the participants felt performance evaluations should be used to point out problems and aid employees in their career directions. Those statements that pointed at using evaluations for rewards or penalties received the worst ratings, showing that the participants felt these should not be the purposes for which job performance evaluations are conducted. The responses to this section given by survey participants coincide with the concepts on which total quality management (TQM) is based, which include the idea of focusing on continual

Table 11.7 Ratings of performance evaluation goals by supervisors

Goals for performance evaluations	Average rank
	(1 = Most important, 5 = Least important)
Make company aware of existing problems	1.65
Provide communication between supervisor and employee	1.80
Show appreciation for employee's efforts	1.81
Give feedback to employee about job performance	1.86
Set goals/give direction for employee to work toward	1.95
Assist in improving employee's skills	2.05
Evaluate processes and materials used on job	2.08
Determine who works in what job	2.13
Determine personnel needs of the company	2.18
Determine employee's training needs	2.19
Personality compatibility evaluations for determining work crews	2.33
Provide documentation of employee's work history	2.51
Award or penalize employees based on performance	2.51
Have employees appraise own strengths, weaknesses	2.65
Basis for promotion	2.76
Provide ranking system to rate employees	2.88
Determine employee's pay	3.41

improvement within one's company. The list of goals with their respective average ratings is shown in Table 11.7.

SURVEY OF CRAFTSMEN'S DUTIES AND ATTRIBUTES

A survey was also developed to gather information about craftsmen. One thousand surveys were mailed to NECA contractors. Two hundred and twenty-nine were returned (22.9 percent response rate). Over 80 percent of respondents classified themselves as either upper management or management.

For the purposes of the survey and this chapter, the term craftsmen is defined as workers who are directed by a supervisor. This study groups four types of workers under the name craftsmen and included material handlers, apprentices, journeymen, and foremen.

Material handlers, who are sometimes called warehousemen, are entry-level workers. They are on the first rung of the ladder that leads to becoming a journeyman. Above the material handlers are the apprentices. These are workers who are studying to become journeymen. When apprentices finish their schooling, they graduate to journeymen. Journeymen who show themselves to be quality workers and possible leaders can be promoted to foremen. Foremen supervise journeymen, apprentices, and material handlers. Depending on the size of the project, a foreman may report to the project supervisor.

Craftsmen Duties

The goal of this section of the craftsmen survey was to determine those duties that are viewed as important parts of a craftsman's job. Respondents were to complete four different sections of the survey: one for foreman duties, one for journeyman duties, one for apprentice duties, and one for material handler (or equivalent position) duties. Each section contained a list of typical, yet specific, job duties associated with each position. Respondents were to rank the duties in order of importance (1 = the most important job duty, 2 = second most important job duty, and so on). The purpose behind ranking the job duties was so respondents could express their view of which duties they felt were most important and which duties they felt were not so important. Singling out the duties that are important is a crucial part of conducting an applicable evaluation of a craftsman's job performance.

Foreman Job Duties

The duties to be ranked in the foreman section of the survey were obtained through interviews with several foremen from different electrical contracting companies. In statistically analyzing the overall average ranking given to each job duty, the following five duties stood out as being ranked significantly higher than the rest:

1. Supervising subcontractors, journeymen, apprentices, material handlers
2. Planning and laying out jobs
3. Maintaining positive working relationships with customers
4. Monitoring safety in the field; conducting regular safety meetings
5. Completing regular reports and paperwork

Because these duties are commonly associated with a supervisory role, this shows that respondents see supervisory functions as the most important aspects of a foreman's job. The list of foreman job duties and each duty's ranking are shown in Table 11.8.

Table 11.8 Ranking of foreman job duties

Job duty	Average rank
	(1 = Most important, 8 = Least important)
Supervising subcontractors, journeymen, apprentices, material handlers	1.92
Planning and laying out jobs	2.03
Maintaining good working relationships with customers	3.88
Monitoring safety in the field; conducting regular safety meetings	3.90
Completing regular reports and paperwork	4.79
Handling claims and change orders	5.92
Handling personnel matters (hiring, firing, evaluating workers)	6.12
Estimating and scheduling	6.30

Journeyman Job Duties

This survey section listed job duties common to journeymen. This list was created through interviews with several journeymen from different electrical contracting companies. Each of the 229 survey participants ranked the list of duties in order of importance. When the average rankings given to each job duty were analyzed, the two duties stood out as receiving rankings significantly higher than the rest:

1. Installation of materials/equipment
2. Following/carrying out instructions

These duties show that respondents see a journeyman's position as one that is focused on labor and output. One journeyman duty stood out as being significantly unimportant to the survey respondents: help with estimating, scheduling, purchasing, and so on. This result may show that this is not seen as a common job duty of a journeyman. The entire list of journeyman job duties and each duty's respective ranking are shown in Table 11.9.

Apprentice Job Duties

The list of duties in the apprentice section of the craftsmen survey was created through interviews with several apprentices from different electrical contracting companies. In statistically analyzing the overall average ranking given to each job

Table 11.9 Ranking of journeyman job duties

Job duty	Average rank
	(1 = Most important, 9 = Least important)
Installation of materials/equipment	1.88
Following/carrying out instructions	2.94
Reading and interpreting drawings	4.12
Practicing and helping enforce safety measures in the field	4.51
Knowledge of codes	4.72
Help instruct and motivate apprentices	5.30
Seeking out shortcuts and/or creative solutions	5.46
Maintaining customer relationships	6.30
Help with estimating, scheduling, purchasing, and so on	8.29

duty, the three following duties stood out as being ranked significantly higher than the rest:

1. Following/carrying out instructions
2. Installation of materials/equipment
3. Practicing safety

These rankings show a similar pattern to that of the journeymen rankings, but with less emphasis on production and more emphasis on practicing. This is logical, since the apprentice position is one of learning. The complete list of apprentice duties and each duty's respective ranking are displayed in Table 11.10.

Material Handler Job Duties

The duties to be ranked in the material handler section of the survey were obtained from interviews with several material handlers from different electrical contracting companies. When the overall average rankings in the list were statistically analyzed, the following two duties stood out as being ranked significantly higher than the rest:

1. Knowledge of materials and tools
2. Following/carrying out instructions

These match the common description of a material handler's (or equivalent position) job: retrieving tools and materials for the journeymen upon their request. The complete list of material handler job duties and the respective rankings are shown in Table 11.11.

Table 11.10 Ranking of apprentice job duties

Job duty	Average rank
	(1 = Most important, 7 = Least important)
Following/carrying out instructions	1.83
Installation of materials/equipment	2.33
Practicing safety	2.95
Good housekeeping	4.51
Knowledge of codes	4.94
Reading and interpreting drawings	5.22
Maintaining customer relationships	5.29

Table 11.11 Ranking of material handler job duties

Job duty	Average rank
	(1 = Most important, 4 = Least important)
Knowledge of materials and tools	1.76
Following/carrying out instructions	1.84
Practicing safety	2.74
Good housekeeping	3.41

Craftsmen Attributes

In this section of the craftsmen survey, respondents were asked to rank lists of attributes found among competent craftsmen. This section contained four lists of attributes, including those for (1) foremen, (2) journeymen, (3) apprentices, and (4) material handlers. Respondents were asked to rank the qualities listed in order of importance (1 = most important attributes, 2 = second most important attributes, and so on). The purpose of having respondents rank the attributes was the same as in the duties' sections. Determining which attributes are important can help in properly evaluating the job performance of a craftsman.

Foreman Attributes

The attributes listed in the foreman section were created through interviews with several foremen from different electrical contracting companies. When the overall average rankings given to each attribute were statistically analyzed, the following four attributes stood out as being ranked significantly higher than the rest:

1. Leadership
2. Organization

3. Work ethic
4. Ability to solve problems

Leadership was given a far better ranking than its counterparts, which shows this is an essential job-related attribute for a foreman. This makes sense, since foremen are typically in charge of one or several crews. These rankings match the supervisory nature of the foreman job duties that received high rankings. The complete list of foreman attributes and the respective ranking are displayed in Table 11.12.

Journeyman Attributes

Attributes to be ranked in the journeyman list were acquired through interviews with several journeymen from different electrical contracting companies. In statistically analyzing the overall average rankings given to the list, the following seven attributes stood out as being ranked significantly higher than the rest:

1. Productivity
2. Work ethic
3. Enthusiasm and high motivation
4. Common sense
5. Mechanical ability
6. Self sufficiency
7. Ability to visualize work

The attributes listed center around production and output. These rankings coincide with the job duties perceived as important for a journeyman. Attributes that stood out as being unimportant were physical capabilities and knowledge of new

Table 11.12 Ranking of foreman attributes

Attribute	Average rank
	(1 = Most important, 9 = Least important)
Leadership	1.85
Organization	4.00
Work ethic	4.00
Ability to solve problems	4.05
Safety awareness	4.78
Technical abilities	4.87
Concern for workers	5.57
Keep up on paperwork	6.29
Knowledge of new materials, equipment, technology	7.69

materials, equipment, and technology. The entire list of journeymen attributes and the respective rankings are shown in Table 11.13.

Apprentice Attributes

The attributes contained in the apprentice list were compiled through interviews with several apprentices from different electrical contracting companies. In statistically analyzing the overall average ranking given to each attribute, the following five attributes stood out as being ranked significantly higher than the rest:

1. Willingness to learn
2. Work ethic
3. Productivity
4. Common sense
5. Mechanical ability

Since the apprentice position is one of a trainee, it makes sense that the job-related attribute ranked as most important was "willingness to learn." The other rankings focus on the productive output of the worker. These rankings coincide with the apprentice job duties that received high rankings.

The complete list of apprentice attributes and the respective rankings are shown in Table 11.14.

Material Handler Attributes

The attributes listed as common to material handlers were compiled through interviews with several material handlers from different electrical contracting

Table 11.13 Ranking of journeyman attributes

Attribute	Average rank
	(1 = Most important, 9 = Least important)
Productivity	2.47
Work ethic	3.18
Enthusiasm and high motivation	3.37
Common sense	4.43
Mechanical ability	4.54
Self sufficiency	5.13
Ability to visualize work	5.33
Physical capabilities	7.13
Knowledge of new materials, equipment, technology	7.89

companies. When the overall average rankings were statistically analyzed, the three following attributes stood out as being ranked significantly higher than the rest:

1. Work ethic
2. Initiative and motivation
3. Common sense

These rankings show that, even though a material handler is the bottom rung on the ladder of workers, it is important for him or her to stay motivated and work hard to move up the ranks. The full list of material handler attributes and the respective rankings are displayed in Table 11.15.

Table 11.14 Ranking of apprentice attributes

Attribute	Average rank
	(1 = Most important, 10 = Least important)
Willingness to learn	1.75
Work ethic	3.22
Productivity	3.86
Common sense	4.50
Mechanical ability	4.61
Able to ask for help	6.25
Ability to visualize work	6.48
Self sufficient	6.60
Physical capabilities	7.75
Good math knowledge	8.10

Table 11.15 Ranking of material handler attributes

Attribute	Average rank
	(1 = Most important, 6 = Least important)
Work ethic	2.33
Initiative and motivation	2.47
Common sense	2.70
Self sufficient	3.80
Looks out for quality	4.62
Physical capabilities	4.79

Goals of Performance Evaluations

The final section of the craftsmen survey asked why construction personnel think job performance evaluations should be conducted. Survey participants were given a list of 17 statements reflecting potential goals of performance evaluations. They were to give each statement a numerical rating indicating its importance as a goal for performance evaluations. The ratings from which respondents could select were 1 (extremely important), 2 (somewhat important), 3 (not too important), 4 (not important at all), and 5 (does not apply).

Analyzing these ratings gave the research team an idea of what industry personnel think job performance evaluations should accomplish. In reviewing the goals that received the best ratings from the survey participants, the research team found that craftsmen had similar opinions to supervisors as they both felt performance evaluations should be used to point out problems, foster communications among the ranks, and aid employees in their career directions. Those statements that pointed at using evaluations for rewards or penalties received the worst ratings, showing that the participants felt these should not be the purposes for which job performance evaluations are conducted. The list of goals with their respective average ratings is shown in Table 11.16.

EVALUATION PROCESS DETAILS

Companies That Conduct Job Evaluations

Fifty-nine out of 193 supervisors' survey respondents indicated that their firms conduct evaluations and 69 out of 229 of the craftsmen survey respondents indicated the same. Thus, approximately 30 percent of both groups responded that their companies conduct job performance evaluations.

Among both groups of respondents, results showed that a company was more likely to conduct job evaluations the more work it performed. That is, the higher the annual dollar amount of work performed, the more likely a company was to conduct evaluations. For example, according to the supervisors' survey, for companies with less than one million dollars in annual volume, fewer than 20 percent perform evaluations; while for companies with $10 million to $20 million in annual volume, more than 60 percent perform evaluations.

In both surveys, the more full-time journeymen the company employed, the more likely it was to conduct job evaluations. In addition, results from both surveys showed that companies located in the Southeastern and South Central regions of the United States were more likely to conduct evaluations. Finally, neither survey showed a correlation between companies that conduct performance evaluations and the type of electrical construction that the companies do.

Table 11.16 Ratings of performance evaluation goals by craftsmen

Goals of performance evaluations	Average rank
	(1 = Most important, 5 = Least important)
Provide communication between supervisor and employee	1.59
Make company aware of existing problems	1.61
Assist in improving employee's skills	1.65
Give feedback to employee about job performance	1.83
Determine employee's training needs	1.94
Show appreciation for employee's efforts	1.96
Set goals/give direction for employee to work toward	2.08
Provide documentation of employee's work history	2.16
Determine personnel needs of the company	2.30
Have employees appraise own strengths, weaknesses	2.40
Determine who works in what job	2.50
Award or penalize employees based on performance	2.56
Basis for promotion	2.63
Evaluate processes and materials used on job	2.69
Personality compatibility evaluations for determining work crews	2.71
Provide ranking system to rate employees	3.09
Determine employee's pay	3.18

Frequency of Evaluations

In companies that evaluate supervisors, 48 percent evaluate performance once a year; 20 percent conduct evaluations every six months; and 19 percent evaluate according to other schedules, such as quarterly or at the end of each project. The remaining 13 percent did not answer this question. Overall, when asked how satisfied they were with the frequency of their evaluation schedules, supervisors' survey respondents who held evaluations once a year were the least satisfied. Those who performed on an "other" schedule (such as after each project or as needed) were the most satisfied. Complete statistics are shown in Table 11.17.

In companies that evaluate craftsmen, 39 percent conduct evaluations once a year; 26 percent every six months; and 35 percent per project, quarterly, or

Table 11.17 Frequency of evaluations according to supervisors survey

Frequency of evaluations	Percentage of responses	Satisfaction with frequency of evaluations		
		Satisfied	Unsatisfied	No response
Once a year	47.5	64.3	25.0	10.7
Every 6 months	20.3	75.0	8.3	16.7
Other	18.6	81.8	18.2	0.0
No response	13.6			

as needed. The percentage of respondents who said their frequency of job performance evaluations was adequate was as follows: 52 percent of those who do annual evaluations; 78 percent of those on a six-month schedule; 60 percent of those on a per-project schedule; and 75 percent of those on other schedules, including "as needed." Overall, those craftsmen survey respondents—like the supervisors' survey respondents—who held evaluations once yearly were the least satisfied with the frequency, and those on a six-month schedule were the most satisfied. Complete statistics are shown in Table 11.18.

Time Spent on Evaluations

Thirty-nine percent of the supervisors' survey respondents indicated that they spent between a half hour and one hour on the process of evaluating an employee, while 20 percent spent between one and two hours. Overall, 92 percent of those spending between one and two hours said this was enough time to conduct the evaluation, while 87 percent of those who spent between a half hour and one hour were satisfied. Complete statistics are included in Table 11.19, which shows that the more time spent with a supervisory employee on his or her job performance evaluation, the more satisfied the survey respondents were.

Thirty-nine percent of the craftsmen survey respondents said they spent less than a half hour; 28 percent spend between a half hour and one hour; 32 percent spend one to two hours; and 1 percent spend over two hours. The percentages of those who indicated satisfaction or dissatisfaction with the time spent are shown in Table 11.20. The results suggest that the ideal length of time to spend with a craftsman employee on a job performance evaluation is between a half hour and one hour.

In this area, a difference is found between the craftsmen and the supervisors' surveys: those responding to the supervisors' survey preferred to spend more time on an evaluations than those responding to the craftsmen survey. One explanation is that, because a supervisor's job is more complex, it may take more time to fully discuss all aspects of the job.

Table 11.18 Frequency of evaluations according to craftsmen survey

Frequency of evaluations	Percentage of responses	Satisfaction with frequency of evaluations		
		Satisfied	Unsatisfied	No response
Once a year	39.1	51.9	22.2	25.9
Every 6 months	26.1	77.8	16.7	5.6
Per project	29.0	60.0	40.0	0.0
Other	5.8	75.0	25.0	0.0

Table 11.19 Length of evaluations according to supervisors survey

Length of evaluations	Percentage of responses	Satisfaction with length of evaluations		
		Satisfied	Unsatisfied	No response
Under 30 minutes	28.8	70.5	23.5	6.0
30 minutes to 1 hour	39.0	87.0	8.7	4.3
1 to 2 hours	20.3	91.7	8.3	0.0
Over 2 hours	5.1	100.0	0.0	0.0
No response	6.8			

Table 11.20 Length of evaluations according to craftsmen survey

Length of evaluations	Percentage of responses	Satisfaction with length of evaluations		
		Satisfied	Unsatisfied	No response
Under 30 minutes	39.1	66.7	25.9	7.4
30 minutes to 1 hour	27.5	94.7	0.0	5.3
1 to 2 hours	31.9	72.7	22.7	4.6
Over 2 hours	1.4	0.0	100.0	0.0

CONDUCTING EVALUATIONS

An important use for performance evaluations is to create employee profiles that track an employee's job performance over time. In order for this profile to provide meaning, a record of each evaluation must be kept. Some sort of score needs to be given to the record so that changes over time can be recognized. These two birds can be felled with one stone: keep track of evaluations on a standard form that provides a scoring system.

Development of the Evaluation Form

An evaluation form is a document on which the results of a job performance evaluation can be recorded. This form should contain a space for the employee's name and job title, the criteria that are used to evaluate the employee, a scoring system of some type, spaces for written comments, and an area that contains information specific to the conditions of the evaluation (date, time started, time finished, evaluator's name and job title, space for the signatures of the evaluator and employee).

There are many ways that an evaluation form can be organized. Your company needs to select a design that best suits your needs. The important things to remember in an evaluation form design are that the form should be easy to use and follow, and should be standardized.

Your company might select a form that encompasses all types of workers, or it might select many similar forms designed to evaluate specific types of workers. Whichever type of form fits your needs, it is important to keep the forms as standard as possible.

Frequency of Evaluations

The surveys performed in this study show that the more often an employee is evaluated, the better the whole evaluation process works. Contractors who indicated that they are happy with their evaluation systems conduct their evaluations at least twice a year. It has also been shown that evaluations work more effectively when the company stays faithful to the evaluation schedule. Making the commitment to conduct performance evaluations regularly shows that you are dedicated to improving your workforce. Listed below are suggestions for evaluation schedules:

- Every six months—Pick two months out of the year (that are six months apart) that will work best for your company. If time is a major consideration, pick months that are not busy.
- Certain points throughout a project—This may vary depending on the length of the project. For example, halfway through and after closeout, after major milestones, or monthly.
- Monthly—Using the first or last week of each month may serve as an easy reminder. For example, a contractor could require that all evaluation reports be turned in by the first of each month.

These are just a few examples of schedules that may work at your company. It is important to select a schedule that will work best for your company. It should also be noted that the simpler your evaluation system, the easier it will be to conduct frequent evaluations.

Tips for Conducting an Actual Evaluation

Perhaps the most challenging part of the performance evaluation process is executing the performance review—the actual meeting between the supervisor and employee to discuss the employee's past work performance. However, this review session is the most important part of the evaluation process. It is where employees find out what they are doing well and what they need to improve.

Studies have shown that if an employee is given recognition for his or her work as well as goals, positive things can happen, including increased productivity, morale, and motivation. Since labor is such an important component of any construction project, these improvements are significant bonuses for contracting companies. Increased productivity means dollars saved on labor; increased morale and motivation means better work put in place. Here are a few tips that can produce successful reviews:

1. Schedule a place and time for the performance review session. This will allow both parties to set aside a special block of time during which the employee's evaluation will be discussed. Scheduling the review in advance will allow both parties to prepare for it (if needed). It can also ensure that enough time will be spent discussing the evaluation.

2. Supervisor completes the evaluation form prior to the review session. Doing this will allow the supervisor to decide his or her own conclusions about the employee's performance. This will also prepare the evaluator to discuss the form in the review session, which can help the review session run smoothly.

3. Employee thinks about own job situation prior to the review session. The review session is a perfect opportunity for an employee to bring up job-related issues and concerns for discussion with his or her supervisor.

4. Supervisor and employee openly discuss evaluation results during review session. Supervisors must be able to provide reasons for the score/ratings given to the employee. While the employee may expect the results, supervisors should be prepared for defensiveness, disappointment, or arguments. It is a good idea for a supervisor to start out the discussion with a few positive comments, as well as have a backup plan in case things get out of hand. Open communication is crucial for making the discussion go smoothly. Both parties need to allow time for questions, comments, and discussion. It is important for both parties to remember that the evaluation should remain as objective as possible. Personal feelings should never enter into the discussion.

5. Be prepared to spend as much time as needed. This study shows that the optimum time to spend on a performance evaluation (filling out forms and discussion) is about one hour. However, each individual is unique, and may require more or less time. The important thing to remember is to allow adequate time for discussion.

6. Decide on action to be taken/set goals for employee to work toward. The evaluation will (hopefully) reveal what the employee is doing well and what the employee needs to improve. It may also reveal that the employee is ready for a promotion, or that he or she should be transferred to another project. Whatever the case, both parties need to agree on what actions to take in improving the employee's job performance. In most cases, this involves setting goals for the employee to accomplish before his or her next performance review. The goals can be specific or general, just as long as the employee knows what he or she needs to work on.

7. Verify that the review session did take place, and that both parties are satisfied with the discussion. This can be as formal as dated signatures from both parties on the evaluation form, or as informal as a handshake. If a disagreement should occur, then the employee can enter into an appeal procedure set up by the company or be referred to the next level of supervision.

8. Remember to conduct performance review sessions regularly. Evaluating job performance regularly can have a positive influence on your company's workforce. Employees will know what is expected of them, and will be better workers because of it. And the better your employees perform, the better your company will be.

Steps for Implementation

Whether you want to implement a new performance evaluation system or are looking to improve an existing one, here are some steps to follow that will help the whole process run more smoothly:

1. Decide on evaluation criteria. Feel free to use the criteria presented in this chapter as a guide. You can use all of the criteria, some of the criteria, or none of the criteria. You can also create your own criteria. The important thing to keep in mind is that the evaluation criteria should match the employee's daily job activities.

2. Develop a scoring system. Feel free to create one that will fit the needs of your company. Keep the scoring system simple, standard, and easy to use and understand. Your company may also want to develop a reward system to accompany the scoring system.

3. Develop an evaluation form. Having a form will allow the evaluation results to be recorded. The form should be as standard as possible, and should include an area for written comments. A form that is short and simple will allow for ease of use.

4. Decide what to do with the completed evaluation forms. A file of completed evaluation forms can make for an excellent employee profile system. There are many ways that such a filing system can be set up; it all depends on each company's individual wants and needs.

5. Decide who evaluates whom. The best evaluators are those who are familiar with the employee's daily job activities. Employees were commonly evaluated by their direct supervisors.

6. Decide how often evaluation review sessions will be conducted. The surveys presented in this chapter show that performance evaluation systems work best if review sessions are held at least twice a year. However, you need to decide what will work best for your company. Whatever you decide, it is important to stick to it and keep it regular.

7. Establish standard guidelines for conducting review sessions. Conducting the review sessions is, perhaps, the most challenging part of the performance review. It will help (especially in the early stages of implementation) if the evaluators have a guide to follow for how to conduct a review session.

8. Announce new system to employees. The whole idea of a performance evaluation system will most likely have better acceptance by the employees if they know it is coming and what to expect. It is a good idea to explain to the employees the reasons why such a system is being implemented, along with an explanation of how it will be implemented. Cooperative employees will make this a successful program in your company.

9. Be flexible. It is rarely easy to implement something completely new. Problems may occur. Be open to suggestions, and be willing to make changes. The patience you have will pay off. Once the system gets rolling, it will reap great benefits to your company.

CONCLUSIONS

Both the supervisors and the craftsmen survey results showed that only about 30 percent of the responding companies conduct job performance evaluations. Results from both surveys revealed that the tendency for a company to conduct evaluations was linked to how much work the company performed per year as well as how many people were employed. However, research has indicated that there are tangible advantages when performance evaluations are conducted in an organization.

The major contribution of this type of study is that it can provide a tool for small companies to get started toward the goal of conducting performance evaluations with their field personnel.

The results of this study were obtained by surveying the state of job performance evaluations of supervisors and craftsmen in electrical construction companies. However, contractors from other trades may also benefit from these results since the different trades in the construction industry share characteristics regarding performance evaluations of supervisors and craftsmen. Contractors from other trades may want to customize the criteria to incorporate the unique elements of their trade.

WEB ADDED VALUE

Evaluations.doc

This file provides sample evaluation forms that contractors can customize for implementation in their companies. Two samples are provided to evaluate supervisors, one to evaluate journeymen, and one to evaluate apprentices.

Web
Added
Value™

This book has free material available for download from the
Web Added Value™ resource center at *www.jrosspub.com*

12

CONTRACT RISK MANAGEMENT

Dr. H. Randolph Thomas, *The Pennsylvania State University*
Amy J. Lesher, *The Pennsylvania State University*
C. Grainger Bowman, *of the law firm Powell, Trachtman, Logan, Carrie, Bowman, and Lombardo*

INTRODUCTION

When a specialty contractor signs a contract to provide construction services, the terms of that contract will be enforced by courts of law even though the contract may contain burdensome terms or provisions that result in harsh consequences. This chapter assists the specialty contractor in several fundamental ways. It focuses attention on selected clauses contained in the general conditions of the contract. Contract clauses are generally of three types: (1) rights and responsibilities—define what the parties are to do and how they relate to other parties and to the project execution plan; (2) entitlements—define the circumstances when the specialty contractor is entitled to additional time or monies; and (3) procedures—define how various administrative aspects of the job, such as changes, payments, notice will be handled; additional information may also be provided. Selected key clauses in each of these categories are discussed. By focusing on these clauses, most of the important risk allocation clauses are covered. Examples of clauses with minimal, intermediate, and high risks are given. In this way, specialty contractors can focus on the key elements of each clause. A project-specific methodology for relating contract risks to project risks and the economic exposure created by the contract and the project is also presented.

PREPARING FOR CONTRACT RISK MANAGEMENT

For any project, it is important to protect the assets of the company. The risks are too numerous, too great, and too detailed to be completed solely by the owner of the company. Therefore, the owner needs to organize a risk management team with each person having responsibility for assessing and managing risks in specific areas. It is incumbent of the owner to know the management responsibilities in the contract. Table 12.1 illustrates the key risk management members who should participate in the analysis of the different contract clauses described in this chapter. The risk management team and their respective roles are described in the following paragraphs.

Specialty Contractor's Owner

Every time a contract is signed, the assets of the company are committed. The owner has the ultimate authority to commit these assets; therefore, the owner must know the risks associated with the project. It is the owner's responsibility to organize a contract risk management team to get advice on the risks. The owner also handles issues related to the allocation of home office overhead, resources, insurance, and other broad issues affecting the company. Risks associated with payments and liquidated damages are the owner's responsibility.

Project Manager

The project manager is responsible for organizing and managing the resources needed to complete the work. The project manager must understand the procedures required by the contract to assure that entitlements due the specialty contractor are not unintentionally forfeited. These include procedures for payment notice, change proposals, submittal reviews, and so on. The project manager must know the risks associated with planning and scheduling the work.

Construction Superintendent

The field or construction superintendent organizes the resources, assures quality, and interfaces directly with other contractors. He must be familiar with the technical specifications and all of the requirements of the work. He must also be familiar with the procedures and other requirements of the general contract conditions.

Table 12.1: Contract risk assignments

Contract clause	Specialty contractor's owner	Project manager	Construction superintendent	Estimator	Scheduler	Insurance agent	Attorney
Indemnification	•					•	•
Coordination of trades		•	•		•		
Schedule acceleration		•	•		•		
Waiver of future claims	•	•					
Concealed conditions			•	•			
No damages for delay				•	•		
Time extensions		•	•	•	•		
Extras and changes	•	•	•				
Pass through	•	•	•				
Pay when paid	•	•					

Estimator

The estimator gives the owner an accurate number with which to finalize the bid. The estimator must understand the requirements of the project fully. Any ambiguities and uncertainties need to be identified and the related risks evaluated.

Scheduler

The scheduler needs to know the order in which the work is to be done and how much flexibility there is in the schedule. The scheduler needs to be able to determine how to keep the work moving forward should there be difficulties in coordinating with other contractors or if there is schedule acceleration.

Insurance Agent

The insurance agent knows how to assess risks and needs to be consulted with respect to the insurance and indemnification requirements on each project to determine if the events required to be covered in the contract are in fact included in the contractor's insurance policy.

Attorney

Although the attorney is probably the least important member of the risk management team, the attorney can provide valuable insight into the understanding of certain contract clauses. The attorney can also be consulted as problems arise during contract performance.

The role of the contract risk management team is to minimize most of the potential mistakes that can be made relative to the project. Its approach should be proactive rather than reactive. In this regard, the team should anticipate what can go wrong and know how the consequences can be mitigated. To do so, there must be effective management of the construction contract.

A contract is a legally binding agreement between two or more parties to exchange something of value, typically monies for construction services. All construction projects are built according to a contract that describes what is to be built, how it is to be built, and when it is to be finished. There are three important aspects about understanding contract risks:

1. Risks are often related, that is, when one event occurs, other events may follow.
2. The likelihood of an event occurring is dependent on the factors surrounding the project, that is, the geographical locale, design, complexity, and number of and working relationships with the par-

ties. Understanding these factors is essential in effectively evaluating contract risks.

3. Assessments are subjective evaluations by the contractor; inaccurate assessments can lead to disastrous economic consequences. There is no analytical way to perform this task.

It is exceedingly important to understand what the contract requires of the specialty contractor. Yet, reading and understanding a construction contract can be a tedious and laborious task for which few contractors have received any formal training. Sadly, many times, specialty contractors do not carefully read the contract. The harsh reality is that if a dispute arises over a particular event, the specialty contractor may find that it has been assigned that risk, and the owner and prime contractor will refuse to pay. If the dispute is taken to arbitration or court, there will be little relief and no sympathy for signing a contract with clearly worded consequences. The only way to avoid or minimize the consequences is to read the contract carefully and make wise risk assessments.

The clauses that are described in this chapter are organized into three groups as follows:

1. Rights and responsibilities clauses
   - Indemnification
   - Coordination of trades/out-of-sequence work
   - Schedule acceleration
   - Waiver of future claims
2. Entitlement clauses
   - Concealed conditions
   - No damages for delay
   - Time extensions
3. Procedural clauses
   - Extras and changes
   - Pass through
   - Pay when paid

RIGHTS AND RESPONSIBILITIES CLAUSES

Indemnification Clause

The indemnification clause contractually shifts the risk of economic loss from one party to another. Through the indemnification clause, the specialty contractor may be required to assume certain liabilities of the owner, prime contractor, or other specialty contractors. The liabilities addressed in indemnification clauses usually arise from accidents resulting in personal injury and property damage and from environmental hazards. These liabilities are usually insurable, thus spreading the risk throughout the industry. However, not all hazards are insurable. Further, some insurance policies will only cover liabilities when the specialty contractor is legally liable, not contractually liable. Therefore, it is important for specialty contractors to understand the coverage limitations of their insurance policy. It may be necessary to purchase additional insurance coverage.

The subcontractor owner, insurance agent, and to a lesser extent the attorney are the key members of the risk management team that should analyze this type of clause. The owner, along with the insurance agent, must review the insurance policy to determine which hazards are covered. For some uninsured hazards, additional insurance can be purchased while other hazards may not be insurable. The uninsurable hazards must be identified. Terms such as *co-insured* need to be defined. The attorney can provide input on the legal limits of the indemnification clause and other implications.

The indemnification clause is a rights and responsibilities clause. It protects the owner, prime contractor, or other named parties against claims for certain losses. Generally, the types of losses are injury or damage to persons and property including the cost of court and attorney fees. The indemnification clause is found in all contract forms, including the contract forms published by the American Institute of Architects (AIA), Engineers Joint Contract Documents Committee (EJCDC), Federal Government, and various standard subcontract forms. However, in many nonstandard forms, the requirements of the clause will vary appreciably.

The indemnification clause is usually complete, and normally, there is no language found elsewhere in the contract to modify the coverage of the clause. However, there can be exceptions; for instance, a contract may modify the indemnification clause for certain hazardous materials. The insurance coverage required by the contract should also be reviewed. Other contract topics that may contain indemnification requirements include subcontractor acts and omissions, sub-subcontractors, patent claims, and safety. Providing specific insurance coverage for the benefit of others is also a form of indemnification. Therefore, the insurance coverage clauses should be carefully read to determine which parties are to be named as additional insured to the policy.

In general, indemnification clauses are disfavored by the courts, and the extent of coverage will be narrowly limited. Some courts have decided that broadly-worded clauses that indemnify parties for their own negligence are against public policy. Other courts have held that to be enforceable, a contract must specifically state that the parties intend to indemnify the protected party even from its own negligence. Yet, other jurisdictions have interpreted indemnification clauses as written.

In reviewing the indemnification clause, the principle issues are as follows (1) What types of economic losses are covered? (2) Does indemnity extend to situations that are not the specialty contractor's fault? (3) What damages are indemnified?

There are three basic forms of the indemnity clause. One form of the clause holds each party responsible for damages only to the extent that the party's own negligent acts caused the injury. Another form limits the assumption of risk to claims due to the conduct or negligence of the subcontractor. However, the contractor is liable for the total damages, even if other parties contributed to the cause. In the broadest form, the contractor agrees to indemnify other persons for claims or damages even if the contractor bears no share of the fault.

How Much Risk Is Assigned to the Specialty Contractor?

Minimal Risk. Generally, minimal risk is assigned to the specialty contractor in the clauses contained in the standard contract forms. The language in these contracts is essentially the same. A typical example is as follows:

> *To the fullest extent permitted by law, the Contractor shall indemnify and hold harmless the Owner, Architect, Architect's consultants, and agents and employees of any of them from and against claims, damages, losses and expenses, including but not limited to attorneys' fees, arising out of or resulting from performance of the Work, provided that such claim, damage, loss or expense is attributable to bodily injury, sickness, disease or death, or to injury to or destruction of tangible property (other than the Work itself) including loss of use resulting therefrom, but only to the extent caused in whole or in part by negligent acts or omissions of the Contractor, a Subcontractor, anyone directly or indirectly employed by them or anyone for whose acts they may be liable, regardless of whether or not such claim, damage, loss or expense is caused in part by a party indemnified hereunder. Such obligation shall not be construed to negate, abridge, or reduce other rights or obligations of indemnity which would otherwise exist as to a party or person described in this Paragraph. The obligations of the Contractor under this Paragraph shall not extend to the liability of the Architect, the Architect's consultants, and agents and employees of any of them arising out of (1) the preparation or approval of maps, drawings, opinions, reports, surveys, Change Orders, designs or specifications, or (2) the giving of or the failure to give directions or instructions by the Architect, the Architect's consultants, and agents and employees of any of them provided such giving failure to give is the primary cause of the injury or damage.*

For this minimal risk clause, the specialty contractor has limited exposure and only for things that are wholly or partly his or her fault. The types of economic losses are insurable.

Intermediate Risk: The following clause requires the subcontractor to pay all costs when there is partial negligence on his or her part. Notice that economic losses are also included.

The subcontractor expressly agrees to save and hold harmless, indemnify and defend the contractor, owner, and the architect or engineer from and against any and all liability, claims, losses, damages, causes of action, costs and expenses, including attorney's fees, arising or allegedly arising from personal injury or the death of any person, including but not limited to employees of the subcontractor, property damage, including loss of use thereof, economic loss, or otherwise, arising or growing out of the work performed by the subcontractor or for the subcontractor's account under this agreement, including any claim or liability arising from any act, error, omission, or negligence of the contractor occurring concurrently with that of the subcontractor or contributing to any loss indemnified hereunder, except for the sole negligence or willful misconduct of the contractor. This indemnity is not intended to extend to any claim arising from the negligence of the architect or engineer related to or arising from the design and/or engineering for the project.

Maximum Risk: The following clause assigns maximum risk to the specialty contractor:

To the maximum extent permitted by law, Subcontractor hereby assumes the entire responsibility and liability for any and all damage (direct or consequential) and injury (including death), of any kind or nature whatsoever, to all persons, whether or not employees of Subcontractor, and to all property and business or businesses, caused by, resulting from, arising out of, or occurring in connection with (i) the Work; (ii) the performance or intended performance of the Work; (iii) the performance or failure to perform the Contract; or (iv) any occurrence which happens in or about the area where the Work is being performed by Subcontractor either directly or though a Subcontractor, or while any of Subcontractor's property, equipment or personnel are in or about said area. Except to the extent, if any, expressly prohibited by law, should any such damage or injury be sustained, suffered, or incurred by Owner, Developer, Architect/Engineer, or Contractor or should any claim for such damage or injury be made or asserted against any of them, including any alleged breach of any statutory duty or obligation on the part of Owner, Developer, Architect/Engineer or Contractor, Subcontractor shall indemnify and hold harmless Owner, Developer, Architect/Engineer and Contractor, their officers, agents, partners, employees, affiliates and subsidiaries (hereafter collectively referred to as indemnitees), of, from and against any and all such damages, injuries, and claims and further, from and against any and all other loss, cost, expense, and liability, including without limitation, legal fees and disbursements, that any Indemnitee may directly or indirectly sustain, suffer or incur as a result of such damages, injuries, and claims; (. . .) In event that any such claim, loss, cost, expense, liability, damage or injury is sustained, suffered, or incurred by or is made, asserted or threatened against Indemnitees, Contractor shall, in addition to all other rights and remedies, have the right to withhold from Subcontractor any payments and to become due to Subcontractor an amount sufficient in Contractor's judgment to protect and indemnify the Indemnitees from and against any such claim, loss, cost, expense, liability, damage, or injury, including legal fees and disbursements; (. . .)

What Is the Likelihood of the Event Occurring?

- Nature of the project: Obviously, certain types of projects are more prone to accidents and mishaps. Projects that are more complex are more prone to accidents. Hazardous materials like asbestos can be found on renovation projects.
- Time of completion: When the schedule is tight, there will likely be more congestion and stacking of trades. All specialty contractors will be trying to finish their work at once. Consequently, these projects pose more risk of accidents and property damage.
- Quality of design: The quality of design is probably not a strong risk factor in indemnification unless the design quality will lead to more changes. Late changes may need to be made in an operating environment and so this factor can interact with a tight schedule to increase the risks.
- Project team: The project team is an important factor. It is necessary that the prime contractor effectively maintains order and discipline and insists on good housekeeping practices. The specialty contractor should talk with other specialty contractors about the team members if there are concerns.

What Is the Economic Exposure?

While the likelihood of accidents and property damage may not be great, the economic exposure can be quite significant. Accidents cost considerable amounts and will affect insurance and workers' compensation premiums for years to come. In addition to the insurable cost of accidents, there are uninsurable costs. These include loss of time to investigate the accident, repairs, loss of labor efficiency, and many other costs. Thus, if the likelihood of incidents occurring is higher than normal, and the contract assigns intermediate or high levels of risk to the specialty contractor, then the economic exposure can be significant. Environmental hazards can also pose significant levels of economic exposure.

Risk Management

Clearly, before the contract is signed, the specialty contractor must carefully review its insurance policy to determine if the risks assigned by the indemnification clause are insurable. After the work has begun, the most effective way to avoid the likelihood of accidents and property damage is a proactive safety program. A strong commitment to safety will prevent accidents and injuries, improve productivity, and contribute to positive morale. Avoid situations in which the work must be accomplished with undue haste because haste leads to shortcuts and increases the likelihood of accidents.

Coordination of Trades/Out-of-Sequence Work

The coordination of the trades clause establishes that the control of the construction premises including the construction sequence and schedule is the responsibility of the owner or the prime contractor. The contractor can change the work sequence and can require the subcontractor to coordinate his or her work with other trades on-site.

The project manager, scheduler, and the construction superintendent are the key team members to assess the risks in coordinating trades and performing out-of-sequence work. It is important to know the likelihood of coordination problems, out-of-sequence work, and changes. The scheduler should provide insight into the flexibility of the schedule. The team members, including the estimator, may suggest ways to minimize the field labor requirements. This may reduce the coordination problems somewhat and add some flexibility to the schedule.

The coordination of trades clause is largely a rights and responsibilities clause, although it is not uncommon to find entitlement language embedded in the clause as well. When included, the entitlement language usually seeks to limit the entitlements of the subcontractor.

How Much Risk Is Assigned to the Specialty Contractor?

Since subcontracts give the contractor the right to control the progress schedule, the main issue imparting risks is whether the specialty contractor is entitled to additional time or money as a result of the schedule disruption or interference by other trades.

Minimal Risk: The American Subcontractors Association (ASA) and the Associated Specialty Contractors, Inc. (ASC) Standard Form of Contract allows reimbursement for additional costs if a general contractor or subcontractor encounters delays caused by others that result in later completion than originally anticipated, in acceleration to attempt to offset previous delays, or in substantial disruption of the orderly progress of the work. In this case, the general contractor and all affected subcontractors are entitled to recover the increased costs of performance including, but not limited to, the additional labor as well as supervision and premium pay required as a result of the delay or acceleration; increased labor and material costs for work performed at a later time than scheduled; cost of tools and equipment used during the extended period; increased insurance and surety bond premiums; losses of productivity due to overtime scheduling, schedule disruption and piecemeal operations; overhead and additional prime costs incurred; and extended overhead for periods of time in excess of those contemplated. This type of clause would present the least risk to the specialty contractor. An example of a minimal risk clause is:

During the progress of the Work, it may be necessary for other contractors and other persons (including personnel of the Authority or Contracting Party) to do work in or about the Work Site. The Authority reserves the right to permit and put such other contractors and such persons to work and to afford them access to the Work Site at such time and under such conditions as does not unreasonably interfere with the Contractor. The Contractor shall prosecute its work continuously and diligently and shall conduct its work so as to minimize interference with such other work. If, notwithstanding Contractor's compliance with the foregoing, such other work interferes with Contractor's performance, then Contractor shall be entitled to an equitable adjustment.

Intermediate Risk: Clauses that impose intermediate risks on the specialty contractor allow time extensions but not monetary damages for schedule disruptions. For example:

Contractor reserves the right to perform or have performed, in and about Subcontractor's Work during the time when Subcontractor is performing its Work hereunder, such other work as Contractor desires. Subcontractor shall make all reasonable effort to perform its Work hereunder in such a manner as will enable such other work to be performed without hindrance from Subcontractor. Subcontractor shall make no claim for damage against Owner or Contractor arising out of such other work or interference to Subcontractor resulting therefrom.

Maximum Risk: The specialty contractor has maximum exposure to risk if the contract does not permit an extension of time and a damage claim for interference of work. Additional risk is imposed on the specialty contractor if the prime contractor places the responsibility for coordination of the subcontract on the subcontractors themselves. For example:

Coordination with other trades will be the responsibility of the subcontractor. Subcontractor shall provide coordination drawings as required. Interference due to lack of complete coordination will be the responsibility of each subcontractor. Potential conflicts, delays and quality concerns must be aggressively pursued by the subcontractor with the other trades before they affect the construction progress.

What Is the Likelihood of the Event Occurring?

- Nature of Project: Projects that have some of the following characteristics create more risk to the specialty contractor: (1) projects with numerous subcontractors; this includes numerous commercial projects executed under multiple prime contracts with specialty contractors; (2) projects where congestion and stacking of trades may be a problem; and (3) work within an operating facility or a complex facility like a hospital or clean room.
- Time of Completion: Projects with tight schedules are more prone to coordination difficulties than others. Other factors that operate with timing to make coordination of trades more difficult include (1) smaller contractors that may be undercapitalized, (2) congestion, and (3) a higher degree of complexity of work.
- Quality of Design: If the design documents are incomplete at the time of bidding, the likelihood of work being put on hold, changes, or design discrepancies will increase. These factors are particularly ominous if the schedule is tight.
- Project Team: The management of the team is important, and there are uncertainties if the team has no experience in working together on prior projects of the type in question. Specialty contractors need to be particularly wary when the contract assigns the responsibility for coordination of trades to the subcontractors.

What Is the Economic Exposure?

The inability to adequately coordinate the trades will lead to delays and disruptions to the work, thereby potentially causing significant increases in labor costs. The level of exposure is related to the experience, management skills of the team, and other factors. If design changes are identified or resolved late in the project, the environment in which the work will be done will be greatly different from the environment that would exist had it been resolved early. There will also be more stacking of trades and congestion. Rework may also increase. An incomplete design can also lead to added material costs, and the time needed for start-up or punch list work can increase considerably. Thus, poor coordination of trades can cause considerable economic exposure.

Risk Management

Easing the impacts of coordination of trades requires a flexible schedule. To maintain flexibility, specialty contractors should strive to minimize the field labor requirements. To a limited extent, this can be accomplished by using preassembled components. Some components can be assembled into larger components in storage areas or off-site. Do as much work as possible away from the installation location. This will minimize crew sizes and labor requirements and will add flexibility to the schedule. Handle materials on a second shift or overtime.

Schedule Acceleration

Most contracts state that the control of the construction premises, including the construction schedule, is the responsibility of the prime contractor. Therefore, the contract allows the prime contractor or other party responsible for the construction schedule to modify it by changing the work sequence, or to direct specialty contractors to accelerate their work in order to complete the project on time.

The risks for schedule acceleration are similar to those related to coordination of trades. The project manager, scheduler, and construction superintendent assess and manage these risks. The typical schedule acceleration clause describes rights and responsibilities related to the schedule. In general, the clause does not discuss entitlement, but sometimes issues of entitlement are found therein. More often, issues of entitlements are found elsewhere in the contract.

The right to modify the schedule is usually found in the time or scheduling sections of a contract. However, there may be relevant language regarding modification or the recovery of damages for such changes in other sections of the contract. Courts uphold the right of the parties to enter into a contract that grants the contractor the right to control the schedule. However, courts have recognized certain exceptions to the strict enforcement of the clause, including interference or hindrance by the owner, delays not contemplated at the time the contract was signed, delays of unreasonable duration, and fraud. Parties must also act in good faith. Thus, arbitrary schedule changes may be viewed as a form of hindrance, bad faith, or even fraud.

How Much Risk Is Assigned to the Specialty Contractor?

The primary risk allocation issue related to the schedule acceleration clause involves entitlement issues. Those that grant entitlement for a broad range of causes for the cost increases pose the least amount of risk. Those that grant no entitlement pose the greatest risk. When the clause does not discuss entitlement issues, other parts of the contract must be consulted. The following clauses containing entitlement language illustrate these points.

Minimal Risk: A typical clause assigning the least risk to the specialty contractor may state:

The subcontractor recognizes that changes will have to be made in the Schedule of Work and agrees to comply with such changes. The Subcontractor shall be entitled to adjustments in the Contract Time and Sum for any schedule changes that materially add to its costs or time of performance.

Intermediate Risk: The risk to the specialty contractor increases when the contract restricts the costs that are recoverable. An example of an intermediate risk clause is:

Subcontractor shall furnish additional labor, expedite deliveries of material and equipment, work overtime and/or a second shift and/or holidays and weekends if directed to do so by Contractor. If the Subcontractor is in default on any provision herein and the Contractor determines such items are required to maintain satisfactory job progress, such additional labor, expediting, overtime, second shift or holiday and weekend work shall be provided by Subcontractor at no cost to the Contractor. If the Subcontractor is not in default of any provision herein, the Contractor shall pay the Subcontractor the actual costs incurred by the Subcontractor to furnish additional labor and to expedite deliveries of materials and equipment, and the actual extra cost over the rate for regular time for overtime work. All such costs shall be substantiated by invoices and time slips checked and approved on a daily basis by Contractor. Subcontractor shall not be entitled to receive any amount for overhead or profit or for any inefficiencies or loss of productivity and shall not assert any claim for overhead or profit or damages due to loss of productivity or inefficiencies.

With the previous clause, the direct costs of acceleration are recoverable as long as the Subcontractor is not in default. However, indirect costs and other damages, including overhead or profit, the premium portion of overtime, insurance, loss of productivity, and inefficiencies are not recoverable.

Maximum Risk: The risk to the specialty contractor is greatest when the contract excludes the recovery of all monetary damages caused by all schedule disruptions. Consider the following:

Subcontractor further acknowledges that as construction progresses it may be necessary for Contractor to change the sequential order and duration of the various activities, including those contemplated by this Agreement to account for unanticipated delays, occurrences and other factors which act to alter Contractor's original schedule. Contractor may require Subcontractor, at no additional cost to Contractor, to prosecute Subcontractor's Work in such sequence as the progress of other subcontractors and the Project Schedule reasonably dictate. It is expressly understood and agreed that the scheduling and sequencing of the Work is an exclusive right of Contractor and that Contractor reserves such right to reasonably reschedule and resequence Subcontractor's Work from time to time as the demands of the Project require without any additional cost or expense to be paid to Subcontractor.

In this clause, the subcontractor is not entitled to recover any costs associated with the changes in the schedule regardless of fault.

What Is the Likelihood of the Event Occurring?

- Nature of the Project: Certain projects are more prone to schedule acceleration than others. The cost of acceleration will likely increase for the following: (1) larger commercial or industrial projects compared to modest-size commercial projects; (2) projects that are larger in size and scope; (3) projects in which accelerated work will be executed within an operating environment; (4) projects in which there are numerous other subcontractors; (5) projects that may yield surprises, as would be the case with the renovation of an old building; (6) projects with fixed deadlines like school buildings and sports arenas; (7) projects with planned outages; and (8) some design/build projects with tight schedules.

- Time of Completion: A red flag for the specialty contractor is if the schedule is extremely tight, and there is minimal likelihood that the owner will extend the time of completion.
- Quality of Design: This factor should be an important yardstick for assessing the likelihood of schedule acceleration. Tight completion schedules will exacerbate this factor. Several important questions are (1) Is the design complete? (2) Has the designer been given sufficient time to finalize plans and specifications (this situation is often reflected in incomplete designs)? (3) Is the design complex requiring considerable coordination? Under these circumstances, the number of changes and response time to submittals and requests for information may increase.
- Project Team: The project team includes the owner, designer, contractor and other subcontractors, governmental agencies, and financial institutions. The specialty contractor needs to be wary when there are many team players unless there is a strong, capable party in charge. Team management is important, and when it is lacking the likelihood of cost and schedule overruns can increase significantly.

What Is the Economic Exposure?

Most of the factors related to coordination of trades apply also to schedule acceleration, except the economic exposure is more acute. The factors that occur include numerous changes, rework, congestion, stacking of trades, out-of-sequence work, overtime, overmanning, just to mention a few. Thus, schedule acceleration is expensive, particularly in labor cost. The loss of labor efficiency for the work can easily reach 25 percent or more. Material and equipment cost will probably increase also. Consequently, the likelihood of schedule acceleration should be evaluated carefully because the economic exposure is great.

Risk Management

There are a number of actions that can be taken to minimize the consequences of schedule acceleration. These are described in detail in the chapter "Managing Schedule Acceleration and Completion," Construction Productivity: A Practical Guide for Building and Electrical Contractors (E. Rojas, Editor, 2008). The recommendations cover the management of labor, material and equipment resources, management strategies, and other helpful practices.

Waiver of Future Claims

A waiver is a knowing and intelligent release of an interest. All contracts contain a provision that the acceptance of the final payment constitutes a waiver of all future claims, except those already identified. Some contracts extend the waiver requirement to progress payments. The waiver precludes the owner and prime contractor from being surprised with a claim after the project is complete. Thus, it is necessary for the specialty contractor to notify the owner and prime contractor as claims arise.

The specialty contractor's owner and project manager should assess these risks. The project manager should manage these risks by maintaining a changes and claims log. This log is particularly important when the waiver is tied to progress payments. The waiver of future claims provision imposes a requirement on the contractor to provide specific information. Thus, this is a rights and responsibilities clause. The specialty contractor should carefully review the notice provisions, as most contracts preclude the specialty contractor from waiting to notify the owner until the end of the project. Beware that notice requirements may appear in multiple clauses of the contract.

Contracts expressly state that the subcontractor's acceptance of the final payment constitutes a release of all claims, and these clauses are generally valid and enforceable. The waiver does not apply to claims that have been identified but not yet resolved. However, the contract may require the subcontractor to obtain written acknowledgment from the contractor in order to maintain the right to pursue the claim. Some contracts state that the acceptance of progress payments is a release of any and all claims covered by that progress payment period. However, most courts will not uphold such a waiver if, at the time of the release, the parties were not aware and should not have been aware of facts upon which another cause of action is based.

How Much Risk Is Assigned to the Specialty Contractor?

Minimal Risk: The ordinary situation is that the waiver is tied to the final payment. This is a normal requirement and poses no additional risk or hardship on the specialty contractor. A typical clause from a subcontract reads:

Acceptance by Subcontractor of the Final Payment shall constitute a release of Owner, Developer and Contractor of and from all liability for all things done or not done or furnished or not furnished in connection with the Work, and for every act, omission, or neglect, if any, relating to or arising out of the Project. Before final payment, Subcontractor shall also execute and deliver a General Release to Contractor naming Owner, Developer and Contractor, said General Release to be in such form as Contractor may provide.

Intermediate Risk: However, when waivers are required for each progress payment, there is more risk to the subcontractor as the prime or owner is inclined to use the waiver provision to avoid payment. The specialty contractor should be cautious about clauses such as the following:

> *(. . .) As a further consideration for the granting of this contract to Subcontractor by Contractor, Subcontractor agrees that as a condition precedent to receiving partial payments from Contractor for Work performed pursuant to this Agreement, Subcontractor shall execute and deliver to Contractor with its request for partial payment as above provided, a full and complete release of all claims and causes of action Subcontractor may have or claim to have against Contractor through the date of the execution of said release, save and except for those claims which Subcontractor shall specifically list on said release and describe in a manner sufficient for Contractor to identify such claim or claims with certainty. (. . .)*

What Is the Likelihood of the Event Occurring?

The principal issue related to waivers is the likelihood of numerous changes. When the changes are numerous, it will become more difficult for the specialty contractor to be certain that all changes and claims have been identified. The likelihood of impact damages will also increase.

What Is the Economic Exposure?

The economic exposure of this clause can vary. The only way to minimize the exposure is to be sure that timely notice is provided and all unresolved claims are listed as required. A claims log is an important control tool.

Risk Management

The most proactive step that can be taken to manage the waiver risk is to maintain thorough on-site logs of changes and claims. Be certain to provide timely notice in strict accordance with the contract, and be cognizant of impact damages on changes. Preserve your rights to impact damages through timely and accurate correspondence.

ENTITLEMENT CLAUSES

Concealed Conditions

The concealed condition clause, also known as the subsurface condition clause or the differing site condition clause, outlines the circumstances in which a specialty contractor will be compensated for unknown or hidden conditions. In general, the subcontractor is entitled to the additional costs resulting from hidden physical conditions that *differ materially* from those *indicated* in the contract documents, or conditions that differ from what could be reasonably expected for the work of the character contemplated in the contract.

The estimator and construction superintendent need to review the contract documents to determine the quality and completeness of information related to potential concealed conditions. The on-site visit prior to bidding is an important component of this review. It needs to be completed with care and diligence. The project manager, construction superintendent, and estimator should make this visit.

The concealed conditions clause is an entitlement clause. It also imposes obligations on the specialty contractor to notify the prime contractor that a condition has been found that may entitle the specialty contractor to additional monies and time.

A procedural clause that should be reviewed is the notice requirements clause. Notice requirements may be found in the changes clause. A rights and responsibilities clause that includes the site visitation and investigation is important. Sometimes, other clauses will limit the contractor's entitlement. Language that does so may be found in a number of places in the contract, including the special or supplementary specifications. The site visitation clause may also contain limiting language. Therefore, any clause containing disclaimer language should also be reviewed.

Common law places the risk of performance that proves to be more expensive than anticipated on the contractor. However, courts have recognized the concealed condition clause as a means to distribute some of the burden when dealing with unforeseen or hidden conditions. A key operative phrase in the clause is "indicated in the contract." The condition that may justify entitlement need not be affirmatively expressed in the contract. Generally, it is sufficient that the contract "indicate" through implication or suggestion that a condition will exist and that a different condition was encountered. Thus, two criteria for recovery are that certain conditions were indicated and actual conditions were found that differed significantly from those indicated.

A third criterion is that the specialty contractor must be justified in relying on the conditions indicated in the contract. Several situations may act to make the specialty contractor unjustified in doing so. The specialty contractor must make a reasonable site visit and note key conditions that are reasonably observable. If no site visit is made or easy observable features are not noted during the visit, then the specialty contractor will not likely recover the added cost.

Specific disclaimers and other clauses may also impose on the specialty contractor the obligation to perform specific evaluation studies and other investigations. Also, a disclaimer may be so specific that despite erroneous indications, the specialty contractor may not be able to recover added cost.

If the contract does not contain a concealed conditions clause, then the owner has chosen to assign all the risk of concealed or hidden conditions to the contractor and subsequently to the specialty contractor. When there is no concealed conditions clause, the specialty contractor must show that the contract contained a positive, material statement of fact that turned out to be inaccurate or that relevant information was withheld. The burden is much more formidable than when there is a concealed conditions clause.

How Much Risk Is Assigned to the Specialty Contractor?

The key issues relative to risks are (1) Is there a concealed conditions clause? (2) With the clause, what aspects of the work are covered? In general, as more components of the work are excluded, the riskier the situation becomes.

Minimal Risk: When minimal risk is assigned to the specialty contractors, the owner assumes the responsibility that concealed conditions will increase the cost of performance. A typical clause reads:

If conditions are encountered at the site which are (1) subsurface or otherwise concealed physical conditions which differ materially from those indicated in the Contract Documents or (2) unknown physical conditions of an unusual nature, which differ materially from those ordinarily found to exist and generally recognized as inherent in construction activities of the character provided for in the Contract Documents, then notice by the observing party shall be given to the other party promptly before conditions are disturbed and in no event later than 21 days after first observance of the conditions. The Architect will promptly investigate such conditions and, if they differ materially and cause an increase or decrease in the Contractor's cost of, or time required for, performance of any part of the Work, will recommend an equitable adjustment in the Contract Sum or Contract Time, or both. If the Architect determines that the conditions at the site are not materially different from those indicated in the Contract Documents and that no change in the terms of the Contract is justified, the Architect shall so notify the Owner and Contractor in writing, stating the reasons. Claims by either party in opposition to such determination must be made within 21 days after the Architect has given notice of the decision.

Intermediate Risk: More risk is assigned to the specialty contractor when the concealed conditions clause contains language that limits the application of the clause. The preceding paragraph covers concealed physical conditions. In the following clause, only work on surfaces is covered.

If Subcontractor deems that surfaces of work to which its work is to be applied or affixed are unsatisfactory or unsuitable or differ from any representation hereof or from good construction practice, written notification of the condition shall be given to (Contractor) within five (5) days of when first detected and before Subcontractor proceeds with further work in connection therewith or takes any remedial action with respect to such condition. If Subcontractor fails to give such written notice and fails to receive a response allowing Subcontractor consideration in connection therewith, no consideration or compensation will be given or allowed to Subcontractor with respect thereto.

Maximum Risk: The maximum risk to the specialty contractor occurs when there is no concealed conditions clause. Without the clause, the contractor bears all the risk of concealed conditions. Contracts that intentionally assign all the risk to the contractor often include harsh language elsewhere that closely relates to concealed conditions. This is usually found in disclaimers and site visitation clauses. The following site visitation/disclaimer clause imposes ominous obligations on the contractor and the specialty contractor.

Subcontractor shall carefully inspect the site of the Work, and shall independently and fully satisfy itself with respect to all surface, subsurface, concealed, exposed and other physical conditions at the site, all general and local conditions, and all other matters which might in any way affect the Contract or its performance. Neither the Owner, Contractor, or the Architect/Engineer has made and shall not make representation as to the nature, existence or location of any conditions, existing installations, or obstacles that may be encountered, whether concealed or exposed, or any general or local conditions which may in any way affect the Contract or its performance, and nothing contained in the Contract Documents shall be construed as such a representation. The Subcontractor shall accept the conditions at the site as they eventually may be found to exist and shall make no claim arising out of any unforeseen or unusual conditions or obstacles of any sort encountered, whether physical or otherwise. The Subcontractor shall satisfy itself as to the accuracy of all grades, elevations, dimensions and locations. In all cases of interconnection of its Work with existing or other work, it shall verify at the site all dimensions relating to such existing or other work. Any errors due to the Subcontractor's failure to so verify all such grades, elevations, locations or dimensions shall be promptly rectified by the Subcontractor without any increase to the Contract Price.

The first sentence sets the tone for this ominous clause when it requires that the subcontractor "independently and fully satisfy itself" of all matters relating to concealed conditions. Unfortunately, this requirement to "independently and fully satisfy itself" is a subtle way of requiring the subcontractor to conduct its own evaluation studies. The clause also disclaims responsibility for a variety of causes. Further, part of the disclaimer language is specific. Thus, it may be difficult for the specialty contractor to avoid the harsh consequences of this clause, especially in the absence of a concealed conditions clause.

What Is the Likelihood of the Event Occurring?

- Nature of Project: The following project types are more likely to experience hidden conditions: (1) projects involving rehabilitation and remodeling; (2) commercial projects and school buildings that may contain hazardous materials such as asbestos; and (3) industrial facilities that may contain PCBs and other substances requiring special handling.
- Time of Completion: The timing of completion is probably not an important factor in concealed conditions except that a tight schedule may have precluded adequate investigations of existing conditions by the owner and designer. The time to bid may also have been short, thus limiting the ability of contractors and subcontractors to assess difficulties in performing the work.
- Quality of Design: The quality of design is probably not an important factor in concealed conditions.
- Project Team: The project team is important from the viewpoint of willingness to acknowledge concealed conditions. The specialty contractor should be cautious of novice owners who may have unrealistic expectations about the construction process.

What Is the Economic Exposure?

The economic exposure caused by concealed or hidden conditions can vary widely. Much depends on the nature and extent of the hidden conditions. Further, if the condition is discovered late in the project, then the economic exposure is greater.

Risk Management

Prior to bidding, the specialty contractor should request all evaluation reports and other investigations made by the designer and consultants. These should be studied carefully as they will contain information not included in the contract documents. Always make a thorough site investigation, and take photographs of conditions at the time of bidding. Not examining reports or conducting prudent site investigations will often preclude the recovery of damages. During construction, be sensitive to concealed conditions and provide timely notice in accordance with the contract. Keep accurate records to support the calculation of damages.

No-Damages-for-Delay

The no-damages-for-delay clause protects the owner or prime contractor from claims for monetary damages resulting from delays and disruptions in the execution of the work. The only remedy to the specialty contractor is a time extension. Typical delays include design errors and omissions, change orders, administrative delays, or interference by the owner, architect, or other contractors.

The construction superintendent, scheduler, and estimator need to determine the likelihood of delays and the impacts they may have on job costs. The scheduler needs to determine the amount of flexibility in the schedule. The estimator and construction superintendent may be able to evaluate alternative ways to minimize the field labor component and shorten the schedule.

The no-damages-for-delay clause is an entitlement clause that is not found in the standard contract forms published by the AIA, EJCDC, Federal Government, and Associated General Contractors (AGC). However, the clause is often included in many nonstandard contract forms, especially subcontract forms. There is no standard for how the clause is written. Therefore, a careful reading of the clause is essential since the wording may limit the application of the clause to only selected types of delays.

Generally the no-damages-for-delay clause is valid and will be enforced; however, in some cases courts have shown some reluctance in enforcing the clause because of its harshness. The extent of coverage of the clause likely will be narrowly defined. However, when the wording of the clause is clear and unambiguous, the clause will be enforced as written.

There are several important issues that are related to the enforcement of the no-damages-for-delay clause. The language of the clause may specify certain types of delays for which the clause applies. When there is no limiting language, the clause will apply to all delays and disruptions that are ordinary, usual, and are contemplated by the contract. Examples of delays that are usual and ordinary include:

- Lack of elevators and stairs
- Lack of temporary light and power
- Increased wages
- Lack of temporary heat
- Bad weather
- Accidents
- Material shortages

- Late material deliveries
- Delayed subcontractor performance
- Rework
- Delays in the approval of shop drawings
- Labor inefficiencies

Examples of extraordinary delays not ordinarily covered by the clause include:

- Delayed site access
- Deliberate delays to the delivery of materials
- Unreasonable work schedule

When the behavior of the owner or prime contractor is reprehensible, courts will not generally enforce the clause. In some jurisdictions, this may require a determination of bad faith and unfair dealings on the part of the owner or prime contractor. In others, a mere determination of active hindrance, interference, or failure to fulfill a contract obligation may be sufficient cause to prevent enforcement of the clause. As can be seen, the exceptions are quite formidable, and, under normal circumstances, the clause will be enforced.

How Much Risk Is Assigned to the Specialty Contractor?

In reviewing a no-damages-for-delay clause, the principal issue is if the clause covers all delays, or if there are some types of delays excluded. Obviously, clauses covering all types of delays create the greatest risk for the specialty contractor.

Minimal Risk: The least risk to the specialty contractor is when the contract does not contain a no-damages-for-delay clause.

Intermediate Risk: The risks to the specialty contractor are intermediate when a clause is included in the contract, but covers only delays of a certain type. For example, language limiting remedies to a time extension caused by the "failure or inability of the city to obtain title to or possession of any land" would allow the contractor to recover monetary damages for delays and disruptions caused by other events. Clearly, a careful reading of the contract is necessary. To illustrate this, consider the following language:

> If, as a result of fire, earthquake, act of God, war, strikes, picketing, boycott, lockouts or other causes beyond the control of Contractor, Contractor considers it inadvisable to proceed with the Work hereunder, then Subcontractor shall, upon receipt of written notice thereof from Contractor, immediately discontinue any further Work until such time as Contractor deems it advis-

able to resume said Work. Subcontractor will resume the Work promptly upon receiving written notice from Contractor to do so. Subcontractor shall not be entitled to any damages or compensation on account of any such cessation of Work as a result of any of the causes aforesaid.

The key phrase is "beyond the control of Contractor." Here, causes that are within or partly within the control of the Contractor, such as poor scheduling or coordination or preferential treatment to other subcontractors, are read to be outside the scope of the clause and damages may be recoverable.

Maximum Risk: The maximum risk to the contractor occurs when the no-damages-for-delay clause applies to all delays regardless of the cause. A typical clause is:

The Contractor agrees to make no claim for damages for delay in the performance of this contract occasioned by any act or omission to act of the City or any of its representatives, and agrees that any such claim shall be fully compensated for by an extension of time to complete performance of the work as provided herein.

This clause is straightforward but harsh nevertheless because it covers all delays for any reason. The specialty contractor should be wary of no-damages-for-delay clauses, as the wording is often confusing and nebulous. Consider the following clause taken from another subcontract:

Should Subcontractor be delayed in the prosecution or completion of the work by the act, neglect or default of Owner, Architect or Contractor, or should "Subcontractor be delayed waiting for materials, if required by this Contract to be furnished by Owner or Contractor, or by damage caused by fire or other casualty for which Subcontractor is not responsible, or by the combined action of the workmen, in no way caused by or resulting from fault or collusion on the part of Subcontractor, or in the event of a lock-out by Contractor, then the time herein fixed for the completion of the work shall be extended the number of days that Subcontractor has been delayed, but no allowance or extension shall be made unless a claim therefore is presented in writing to Contractor within 48 hours of the commencement of such delay, and under no circumstances shall the time of completion be extended to a date which will prevent Contractor from completing the entire project within the time allowed Contractor by Owner for such completion. No claims for additional compensation or damages for delays, whether caused in whole or in part by any conduct on the part of

Contractor, including, but not limited to, conduct amounting to a breach of this Agreement, or delays by other subcontractors or Owner, shall be recoverable from Contractor, and the above-mentioned extension of time for completion shall be the sole remedy of Subcontractor, provided, however, that in the event Contractor obtains additional compensation from Owner on account of such delays, Subcontractor shall be entitled to such portion of the additional compensation so received by Contractor from Owner as is equitable under all of the circumstances."

This clause contains limiting language in three locations. At first, it would appear that there is some room for recovering monetary damages; however, the last phrase detailing delays caused in whole or in part by the Contractor, other Subcontractors, and Owner would seem to make the clause inclusive to all types of delays.

What Is the Likelihood of the Event Occurring?

- Nature of Project: The type of project would not seem to be a strong factor in determining the risk associated with the no-damages-for-delay clause, except for the following: (1) more complex projects; (2) projects in which there is owner-procured equipment to be installed; and (3) projects that require approvals from governmental authorities.
- Time of Completion: Obviously, tight time schedules are a concern from the delay standpoint because (1) projects requiring close coordination with owners and designers may lead to delays, (2) interfacing with outside agencies may be slow, and (3) the more parties involved, the greater is the likelihood that someone will be late.
- Quality of Design: Delays caused by late design, changes, or rework are usual and ordinary and fall within the scope of the general no-damages-for-delay clause. Therefore, if late design, changes, and rework are likely, the specialty contractor needs to be wary because recovering delay damages will be improbable.
- Project Team: A team that has worked together on past projects will probably experience less negative impacts than newly formed teams. Be wary of projects in which the owner insists on time-consuming approval procedures for changes and submittals.

What Is the Economic Exposure?

The economic exposure depends on whether the issue is delays or disruptions. The impact of disruptions can be quite severe. The same exposure exists as it would for schedule acceleration. Short-term delays of one day or one week would have the same impact on the work as disruptions. Such delays can cause the work to be out of sequence, which can lead to significant losses of labor efficiency. Longer delays will mean that when the work is finally completed, the work environment will have changed. There will be stacking of trades, congestion, and other factors present. The work may be completed on an overtime schedule, or overmanning may be necessary to maintain the schedule. The most common long-term delays are failure to have site access in a timely manner or a project suspension. In most circumstances, the economic exposure of long-term delays is less than short-term delays.

Risk Management

Minimizing the field labor component of the installation and maintaining flexibility in the schedule will reduce the economic exposure to delays and disruptions. Also, avoid mobilizing equipment and manpower too early.

Time Extensions

Contracts require that the work be performed within the contract time period. However, contracts usually list the circumstances when the contractor will be unavoidably delayed and a time extension will be granted. The clause is often called the force majeure clause.

The project manager and construction superintendent are primarily responsible for evaluating the risks related to time extensions. The estimator and scheduler also have a role in determining lead times of material and equipment and incorporating these factors into the schedule.

The time extension clause details the circumstances when a time extension will be granted. Therefore, it is an entitlement clause. Language related to time extensions may be found in the sections on time of performance, time, delays, and no-damages-for delay clauses. References to time extensions within some contracts are found in the claims section for weather.

By agreeing to perform the work the contractor assumes the risk of most types of delays. Time extensions will be granted only for those events listed in the contract.

How Much Risk Is Assigned to the Specialty Contractor?

The principal issue related to time extensions is what events will be granted a time extension. The fewer events there are defined, the greater the risk to the subcontractor.

Minimal Risk: Contracts that grant an extension of time for ordinary and usual events that lead to delays impose the least risk to the specialty contractor. An example from a standard contract form follows:

The Subcontractor shall not be liable or responsible for delays or costs in completion of the work resulting from or caused by occurrences beyond its control including without limitation changes ordered in the work and other delays resulting from actions or inactions of others; acts of God; floods; fire, explosions or other casualty losses; strikes, boycotts or other labor disputes; lockouts; and acts of the Government.

This clause recognizes multiple types of delay including delays or changes caused by the owner, natural disasters, casualties, and labor problems. A key phrase is "beyond its control," which extends the coverage of the clause beyond those events listed.

Intermediate Risk: Clauses that limit the events for time extensions are more risky to the specialty contractor. In reading the clause, events not listed are intended to be excluded.

Should Subcontractor be obstructed or delayed in the commencement, prosecution or completion of the Work, without fault on its part, by reason of: failure to act, direction, order, neglect, delay or default of the Developer, the Architect/Engineer, the Contractor, or any Other Subcontractor employed upon the Project; by changes in the Work; fire, lightning, earthquake, enemy action, act of God or similar catastrophe; by Government restrictions in respect to materials or labor; or by an industry-wide strike beyond Subcontractor's reasonable control, then Subcontractor shall be entitled to an extension of time to perform the Work which shall be equal to the time lost by reason of any or all of the causes aforesaid, but no claim for extension of time on account of delay shall be allowed unless a claim in writing therefore is presented to Contractor with reasonable diligence but in any event not later than ten (10) days after the commencement of such claimed delay. Except for the causes specifically listed above, no other cause or causes of delay shall give rise to an extension of time to perform the Work. (. . .)

In this clause, material delivery delays not associated with government restrictions or industry-wide strikes will not result in a time extension. Equipment delivery delays are also excluded.

Maximum Risk: Clauses that shift the most risk to the contractor either do not allow time extensions for any reason or severely limit the scope for which they will be granted. In the following clause, the contractor is allowed a time extension for a temporary suspension of work as ordered by the engineer, but not for weather delays. Since no other causes are listed as excused delays, the contractor assumes the risk for any other delays.

> *No payment or compensations shall be made to the Contractor for damages because of hindrance or delay from any cause, whether such delays be avoidable or unavoidable. Whenever the Engineer orders a temporary suspension of work, the Contractor shall be entitled to an extension of time equal to the duration of the suspension. No extension of time will be granted due to adverse weather conditions such as rain.*

What Is the Likelihood of the Event Occurring?

- Nature of Project: The more complex the project is the higher the likelihood for delays. Projects requiring government or authority approvals can also be troublesome. When the owner provides material or equipment, delays may occur.
- Time of Completion: Projects with a tight time schedule will be more prone to delays if the projects were designed or planned in haste.
- Quality of Design: This factor can interact with the time schedule to create significant problems.
- Project Team: Be sensitive to requirements for lengthy or burdensome approval processes.

What Is the Economic Exposure?

The main economic exposure issue related specifically to time extensions is the amount of liquidated damages. Also, determine if the liquidated damages are tied to substantial completion or final completion. The latter creates considerably more exposure.

Risk Management

The project manager and the construction superintendent need to be vigilant to the events that may lead to a time extension. Knowledge of the written requirements for notice is also important. The scheduler plays an important role in the assessment of unexpected events. Therefore, it is important to maintain a current schedule and to have accurate reporting of the field progress. Off-site assemblies and aggressive procurement practices will sometimes minimize the need for extensions of time. Aggressively pursue work as soon as there is a sufficient back-log of work to be performed efficiently.

PROCEDURAL CLAUSES

Extras and Changes

All construction contracts contain a changes clause. Typical of what is covered by the clause is any change in the subcontract work within the general scope of the subcontract including a change in the drawings, specifications, or technical requirements of the subcontract and/or a change in the schedule of work affecting the performance of the subcontract.

Specialty contractor's owner, project manager, and construction superintendent are the key team members. Each needs to be fully aware of the drain on company resources that a project with a large number of changes can have. The notice requirements and change approval procedure need to be fully understood. The likelihood of delays in the approval of changes is an important factor. The pricing of changes should include the estimator.

The changes clause usually found in standard contract forms is generally limited to a discussion of rights and procedures for issuing a change and procedures for arriving at an equitable adjustment for the change. The clause also states that the time or cost of performance cannot be changed without a written change order. However, nonstandard contract forms may also include language related to entitlements. Still other clauses have been found to lack discussions of both entitlement and procedures. Thus, it is important that the changes clause be read carefully and, when lacking certain information, other parts of the contract need to be consulted.

The changes clause may include procedures that are material or essential features of the contract, and these procedures must be followed for specialty contractors to preserve their rights to recover additional costs. Thus, the contract requires affirmative action by both parties.

When the changes clause is primarily a rights, responsibilities, and procedures clause, it will be necessary to examine other parts of the contract for entitlement issues. One exception is that many changes clauses discuss the specialty contractor's entitlement to overhead and profit on changed work. Other procedural language requiring notice may be found elsewhere.

Contracts will be enforced as written. Perhaps the most difficult area for specialty contractors is when the work is performed in advance of a written authorization. When the owner or prime contractor consistently follows the provisions of the contract and insists that the change directive be in writing, the specialty contractor will not likely recover the costs of changes when no written directive is issued. Many contractors have learned this harsh reality.

There are a number of ways an owner or prime contractor can, through their actions, ignore the requirement for a written directive. The criteria for establishing payment for extras has been set forth by the courts as:

- The work was outside the scope of the contract promises
- The extra items were ordered by the owner
- The owner agreed to pay extra, either by his words or conduct
- The extras were not furnished by the contractor as his voluntary act
- The extra items were not rendered necessary by any fault of the contractor

To recover the additional costs, all five criteria must be satisfied.

How Much Risk Is Assigned to the Specialty Contractor?

Minimal Risk: Contracts with the least likelihood of risk will acknowledge changes, establish a clear criterion for notification and approval, and detail how to calculate additional costs. Such clause usually details responsibilities and procedures. Limitations on the percentage allowed for overhead and profit may be the only entitlement issue discussed. Thus, this type of clause leaves the principal issues of entitlement to other parts of the contract.

Intermediate Risk: The risk to the specialty contractor increases somewhat when the clause imposes additional requirements such as (1) procedural requirements that request a definitive acknowledgment of the change from the contractor, await certain authorization directives, or other requirements beyond the normal notice of the need for a change, (2) limitations on allowable percentages for overhead and profit, and (3) clauses that discuss issues of entitlement thus raising the possibility that the clause is inconsistent and uncoordinated with other parts of the contract. An example of the first point is illustrated by the following provision:

If Subcontractor fails to give such written notice and fails to receive a response allowing Subcontractor consideration in connection therewith, no consideration or compensation will be given or allowed to Subcontractor with respect thereto.

While the nature of the response allowing consideration is not clear, the clause does seem to impose an obligation on the subcontractor to await some acknowledgment from the contractor. Such language should be clarified or negotiated out of the contract. Consider the following example:

If the Subcontractor without obtaining prior written approval of the contractor initiates any substitution, deviation, modification, revision, or change in the work to be performed under the subcontract agreement the subcontractor shall be liable for the entire cost thereof. (. . .)

Maximum Risk: Clauses that state that a subcontractor can recover costs only when the prime contractor recovers the monies from the owner place the specialty contractor in a precarious position, especially if the change is ordered by the contractor, but was not preapproved by the owner. In legal vernacular, this situation is referred to as a condition precedent, that is, one condition or event must occur (payment to the prime) before a second event occurs (payment to the subcontractor). Additional risk is also imposed if a method for determining the modified cost is not included in the contract. For example:

Contractor reserves the right to review and validate Subcontractor's proposed cost and/or schedule modification and to determine compliance with Subcontract terms. If it is determined by Contractor, in its sole and absolute discretion, that the cost proposed by Subcontractor is not in compliance or acceptable, Contractor will notify Subcontractor in writing. This notification shall specify all discrepancies and provide an appropriate time requirement for Subcontractor's response. Subcontractor shall respond in writing to justify all discrepancies within the time allotted.

In this case, the contractor has "sole and absolute" discretion in determining the acceptability of the subcontractor's change request.

When there is a condition precedent for payment and the contract contains language that encourages the subcontractor to proceed with work without undue delay, the specialty contractor needs to proceed with caution. The specialty contractor may act in good faith and proceed with the work, but may learn later that recovery of additional costs is precluded. An example of language that should be viewed with caution follows. Observe that the notice period is especially short, only three days.

Contractor may at any time, by written order and without notice to surety, make changes in the work herein contracted for and Subcontractor shall proceed with the work as directed. If said changes cause an increase or decrease in the cost of performance or in the time required for performance, an equitable adjustment shall be made and this Subcontract shall be modified in writing accordingly. Subcontractor shall provide notice of any alleged change to the work or other claim directed or caused by Contractor within three (3) days of the occurrence giving rise to the alleged change or claim or such claim shall be conclusively waived. Subcontractor's entitlement to an adjustment in the cost or time required for performance of a change in the work directed or caused by the Owner shall be governed by the terms of the Contract Documents. Nothing herein contained shall excuse the Subcontractor from proceeding with the prosecution of the work as changed.

What Is the Likelihood of the Event Occurring?

- Nature of Project: Certain types of projects are more prone to changes than others. These include (1) industrial facilities, (2) wastewater treatment plants, (3) hospitals, and (4) other complex structures. Each of these types of facilities can experience more than the normal number of changes. Simpler structures like warehouses and smaller office buildings are less prone to changes.
- Timing of Completion: When projects are on a tight schedule, the problems with changes are more pronounced. There is more pressure to begin executing changes without written authorization just to keep the job moving. The timing of responses, submittal reviews, requests for information, and so forth, can become acute.
- Quality of Design: The quality of the design documents may be less than desirable, especially when schedules are tight. There will likely be numerous changes or rework when the construction schedule significantly overlaps the design schedule. When the design is less than 50 to 60 percent complete when construction begins, numerous changes are highly likely.
- Project Team: The team management can deteriorate quickly, especially when there are cost overruns and schedule delays. At the outset, specialty contractors need to evaluate how likely this is to happen. Consider the following: (1) past experiences with the team members is an important yardstick; (2) be wary when team members seem uncompromising; a situation that can occur is when the need for changes is not acknowledged, and specialty contractors are directed to complete the work to the owner's satisfaction using their own ingenuity; and (3) the owner is potentially underfinanced or lacks contingency monies.

What Is the Economic Exposure?

The economic exposure of the changes clause is usually inconsequential because it is typically a procedures clause. When there is a burdensome approval process and other requirements, the economic impact will be felt in coordination of trades, schedule acceleration, and no-damages-for-delay clauses.

Risk Management

In managing changes and extras it is important to maintain a complete, accurate, and up-to-date log. In pricing changes, be sure to include all elements of cost such as equipment and tools. The cost of on-site overhead should be considered a direct cost of the work and is not included in any overhead limitation unless specifically stated so in the contract. Overhead limitations refer to home office overhead. Be cautious in performing work without a written change directive. It is important to follow the provisions of the contract completely, even when the owner and prime contractor show tendencies toward ignoring those provisions.

Consider tracking the costs of changes separately, particularly when the owner or prime contractor refuses to acknowledge that the work is a change. If impact costs are claimed later, these records will be invaluable to supporting the request for additional cost.

Pass-Through

Pass-through clauses bind specialty contractors to the provisions of the prime contract that affect their work. The main purpose is to require the specialty contractor to be bound by the same performance standards as the prime contractor even though these obligations are not expressively contained in the subcontract.

The specialty contractor's owner, project manager, and construction superintendent will need to be involved in assessing the risks in the pass-through clause. The pass-through clause is a procedural clause and generally does not contain entitlement language. Clauses that contain pass-through language may be titled (1) Mutual Rights and Responsibilities, (2) Compliance with Contract Documents, (3) General Scope, or (4) Work to be Performed.

The pass-through provisions are generally enforceable, and courts recognize that pass-through clauses do not expand the specialty contractor's scope of work. However, interpretation of the clause may vary with jurisdiction. In federal projects, the courts have held that general pass-through clauses only pertain to the definition of the work, not the legal terms and conditions or the administrative procedures in the prime contract. These terms are defined by the subcontract.

Many state courts, on the other hand, do not follow this rule. In these jurisdictions, the conditions and administrative procedures as well as the work obligations and standards within the sections of the prime contract documents are applicable to the subcontractor and become binding through the clause.

How Much Risk Is Assigned to the Specialty Contractor?

Pass-through clauses may bind the specialty contractor to standards included in the prime agreement that may be more restrictive than the subcontract.

Furthermore, the specialty contractor may be agreeing to provisions or standards to which he or she has not even been informed. Therefore, it is important to review all contract documents. The principal elements of risk in the pass-through clause are (1) determining which party will prevail in the event of a discrepancy between the prime contract and subcontract and (2) determining if the specialty contractor has the same reciprocal rights as the prime contractor.

Minimal Risk: The danger of being obligated to unknown standards is minimal when the specialty contractor is only bound to the documents that the prime contractor provides.

"The Contract Documents which are binding on the Subcontractor are as set forth in Article 16.5. Upon the Subcontractor's request the Contractor shall furnish a copy of any part of these documents. These shall not be binding on the subcontractor unless such documents are furnished to the subcontractor.

In the event of a conflict between this Agreement and the Contract Documents, this Agreement shall govern, (. . .)

The order-of-precedence is that the subcontract provisions apply in lieu of the prime contract provisions. To balance the risk assigned by the clause, the specialty contractor should try to create reciprocal rights so that he can also receive the same benefits created by the prime contract. The following clause creates reciprocal rights for the specialty contractor.

"The Contractor and Subcontractor shall be mutually bound by the terms of this Agreement and, to the extent that provisions of the Prime Contract apply to the Work of the Subcontractor, the Contractor shall assume toward the Subcontractor all obligations and responsibilities that the Owner, under the Prime Contract, assumes toward the Contractor, and the Subcontractor shall assume toward the Contractor all obligations and responsibilities which the Contractor, under the Prime Contract, assumes toward the Owner and the Architect. The Contractor shall have the benefit of all rights, remedies and redress against the Subcontractor which the Owner, under the Prime Contract, has against the Contractor, and the Subcontractor shall have the benefit of all rights, remedies and redress against the Contractor which the Contractor, under the Prime Contract, has against the Owner, insofar as applicable to this Subcontract. Where a Provision of the Prime Contract is inconsistent with a provision of this Agreement, this Agreement shall govern.

Intermediate Risk: The following language assigns some additional risk to specialty contractors because now they must read and comprehend the prime contract as well as the subcontract.

The Subcontractor is bound to the Contractor by the same terms and conditions by which the Contractor is bound to the owner under the contract.

Maximum Risk: Since the contractor is responsible for the entire project and site coordination, he or she may not want the subcontractors to have the same rights and remedies as the prime has against the owner. An example of a clause that binds the specialty contractor to a stricter standard follows:

Subcontractor shall be bound by the terms of the Agreement between Owner and Contractor and all documents incorporated therein, including without limitation, the General and Special Conditions, and assumes toward the Contractor, with respect to the Subcontractor's Work, all of the obligations and responsibilities that the Contractor, by the Agreement between Owner and Contractor has assumed toward the Owner. In the event of conflict or inconsistency between this Subcontract and the Agreement between Owner and Contractor and all documents incorporated herein, Subcontractor shall be bound by the provisions, terms or conditions of which impose the greater duty or obligation.

In this clause, the subcontractor is bound by the provision that is the most stringent. Also, the subcontractor does not have reciprocal rights.

What Is the Likelihood of the Event Occurring?

- Nature of the Project: Complex projects that have extensive contract documents and other requirements are more likely to have difficulties with the pass-through clause.
- Time of Completion: This is probably not a strong factor in assessing risks unless the short time of completion means that the contract documents have not been suitably coordinated. The specialty contractor may not have adequate time to review both the prime and subcontract provisions, especially when the documents are extensive and complex.
- Quality of Design: If the design is incomplete, then the other contract documents are also likely to be incomplete and uncoordinated. The specialty contractor may not have adequate time to review both the prime and subcontract provisions, especially when the documents are extensive and complex.

- Project Team: The team members are not likely to be a strong risk factor.

What Is the Economic Exposure?

Since the pass-through clause is a procedural clause, there is limited economic exposure. The dangers of oppressive prime contract provisions will be felt in other clauses.

Risk Management

Managing high-risk pass-through clauses is difficult at best. Considerable time and energy must be spent in reviewing and reconciling the prime contract and subcontract. The outcome of this examination will affect many of the clauses covered in this chapter.

Pay When (If) Paid

The pay-when-paid clause is used so the prime contractor does not incur a liability for payment until it has received the payment from the owner. The clause relieves some of the potential cash flow problems of the prime contractor and shifts these burdens to the specialty contractor.

The specialty contractor's owner and project manager are involved with payments and should be responsible for the risks associated with the pay-when-paid clause. They should assess the likelihood of the owner not making timely payments.

The pay-when-paid clause is a procedural clause because it establishes how the payment process will be handled. If a pay-if-paid clause is used, the prime contractor is not obligated to pay the specialty contractor unless payment is received from the owner. The clause then becomes an entitlement clause. The pay-when-paid clause is usually found in the Payment or Progress Payments sections of the contract.

Generally courts interpret the pay-when-paid provisions to establish the time frame for payment to the specialty contractor. Many courts seem to be reluctant to construe the clause as establishing the owner's payment as a *condition precedent* to the contractor's obligation to pay the specialty contractor. The term *condition precedent* means that if the owner refuses to pay the prime contractor, then the specialty contractor is not entitled to payment from the prime. This is the essence of the pay-if-paid clause. New York and California courts have held that the pay-if-paid clause is unenforceable because it is against public policy. Several other states have statutes limiting the enforceability of the clause. However, a minority of jurisdictions will strictly enforce the clause if the language expressively creates a condition precedent to payment. Particular caution should be used in the states that distinguish between the words *when* and *if*.

How Much Risk Is Assigned to the Specialty Contractor?

The main issue of the pay-when (if)-paid clause is whether the clause creates a *condition precedent* to the obligation for payment, or whether it only limits the time period during which a contractor may delay payment to the subcontractor.

Minimal Risk: The least risk to the specialty contractor occurs when the contract does not contain a pay-when-paid or pay-if-paid clause. Consider this example:

> *Progress payments to the Subcontractor for satisfactory performance of the Subcontract Work shall be made no later than seven (7) calendar days after receipt by the Contractor of payment from the Owner for such Subcontract Work. If payment from the Owner for such Subcontract Work is not received through no fault of the Subcontractor, the Contractor will make payment to the Subcontractor within a reasonable time for the Subcontract Work satisfactorily performed.*

Intermediate Risk: Additional risk is assigned to the specialty contractor if the clause defines the time for payment as well as limits the circumstances for payment.

Maximum Risk: The full risk of nonpayment by the owner is transferred to the specialty contractor if the contract expressively creates a condition precedent to the contractor's obligation to pay. If the owner does not pay, for whatever reason, the contractor is not obligated to provide payment to the specialty contractor. The following clause creates a *condition precedent* to payment. Pay special attention to the last sentence of the paragraph:

> *Provided Subcontractor's rate of progress and general performance are satisfactory to the Contractor, and provided the Subcontractor is in full compliance with the Subcontract Documents, the Contractor will make partial payment to the Subcontractor in an amount equal to 90% of the estimated value of work and materials incorporated in the construction and of materials delivered to the Project site and suitably stored by the Subcontractor, to the extent of Subcontractor's interest in the amounts allowed thereon and paid to Contractor by the Owner less the aggregate of previous payments. Payment will be made by check mailed to the Subcontractor within*

seven (7) working days of receipt of collected funds paid to the Contractor by the Owner under the Prime Contract. Unless otherwise required by law, final payment will be made within thirty (30) days after the work called for hereunder has been completed by the Subcontractor to the satisfaction of the Owner and the Contractor has received from the Owner written acceptance thereof together with payment in full for Subcontractor portion of the work. Final payment is further subject to Contractor's determination that all of the terms, conditions, requirements, and covenants of the Subcontract Document have been well and truly met and discharged by Subcontractor. Payment to the Subcontractor is acknowledged and agreed by the Subcontractor to be expressly conditioned upon the Contractor's receipt of payment from the Owner.

The word *conditioned* or the phrase *condition precedent* should serve as a red flag that the clause is probably a pay-if-paid clause.

What Is the Likelihood of the Event Occurring?

The likelihood that the owner will not pay is an important project risk. Owners who may experience cash flow problems are a particular concern.

What Is the Economic Exposure?

Naturally, the economic exposure for not getting paid can be quite severe.

Risk Management

Discussions with other contractors and specialty contractors may reveal a pattern of prior non- or late payments by the owner or prime contractor. Also, many contracts contain provisions allowing contractors to inquire about the owner's ability to pay for the project or the availability of contingency monies.

CONCLUSIONS

When harsh contract language is encountered, and high risks and economic exposure are possible, then the specialty contractor has several options. First, the specialty contractor should try to negotiate more favorable contract language. However, this may not be possible. The prime contractor may not be willing to negotiate, or the prime may be willing to acquiesce to some less stringent provisions but not to the more stringent ones. The prime's position is usually that there is always another subcontractor who will sign the agreement.

A second option is to not sign the contract and forego the work. In extreme cases, this may be the best option; however, in most instances, it is not a viable option. Thus, the specialty contractor is in the position of most likely having to sign the contract with few substantive revisions. So what does the specialty contractor do when he or she enters into a risky agreement? Do nothing and hope for the best? Fortunately, there are proactive steps that can be taken to limit the contract risks.

Prior to preparing the bid, you should (1) obtain a complete copy of your and the prime's contract; (2) qualify your bid to exclude uninsurable risks and lien waivers; (3) qualify your bid to exclude other items such as red tag or installation charges, consumption charges, temporary power, and cleanup by others unless specifically authorized; (4) talk to your insurance representative and price in your bid the cost of any additional insurance; and (5) determine the track record of the general contractor, owner, and designer.

During pre-construction, one of the important tasks is negotiation. At this stage you should (1) identify the clauses on which one can be flexible and those that are intolerable; (2) strike phrases that give the general contractor discretionary approval, such as "will pay after scrutiny"; (3) consult with your attorney when proposing alternative language; and (4) recognize that even with public entities there may be some flexibility.

During construction you should (1) start work only after you have a signed contract, (2) do your own cleanup, (3) consider site supervision as a direct expense on change work, (4) manage schedule acceleration, (5) include the time impacts and request time extensions for change orders, (6) keep precise daily logs, (7) follow the notice requirements stated in the contract, (8) review minute meetings carefully, and (9) be proactive in getting the shop drawings submitted and approved early.

Finally, during project closeout, you should (1) hold your lien waiver until you can exchange it for your final payment when appropriate and (2) insist on one and only one punch list.

BIBLIOGRAPHY

Ballard, G., and Howell, G. (1998). "Introduction to Lean Construction." *Seminar Proceedings*, Lean Construction Institute.

Ballard, G., Howell, G., and Kartam, S. (1994). "Redesigning Job Site Planning Systems." Conference on Computing in Civil Engineering, ASCE, Washington, D.C.

Beer, M., Eisenstat, R. A., and Spector, B. (1990). "Why Change Programs Don't Produce Change." *Harvard Business Review*, November–December, 158–166.

Bell, L. C., and Stukhart, G. (1987). "Cost and Benefits of Materials Management Systems." *Journal of Construction Engineering and Management*, ASCE 113 (2), 222–234.

———(1986). "Attributes of Materials Management Systems." *Journal of Construction Engineering and Management*, ASCE 112 (1), 14–21.

Construction Industry Institute. (1990). "Total Quality Management: The Competitive Edge." The University of Texas–Austin.

———(1988). "Project Materials Management Handbook." The University of Texas–Austin.

Construction Industry Cost Effectiveness (CICE) (1983). "More Construction for the Money." The Business Roundtable, New York.

Crosby, P. B. (1979). *Quality Is Free: The Art of Making Quality Certain*. New York: Penguin.

———(1984). *Quality without Tears: The Art of Hassle-Free Management.* New York: McGraw-Hill.

Dahlback, O. (1991). "Accident-Proneness and Risk-Taking." *Personality and Individual Differences* 12 (1): 79–85.

Deming, W. E. (1986). *Out of the Crisis.* Cambridge: Massachusetts Institute of Technology.

Drucker, P. (1974). *Management: Tasks, Responsibilities, Practices.* New York: Harper & Row.

Echeverry, D., Ibbs, W. C., and Kim, S. (1990). "Sequencing Knowledge for Construction Scheduling." *Journal of Construction Engineering and Management,* ASCE 116 (3), 118–130.

Fails Management Institute (1994). *Partnering: A Progress Report.* Raleigh, NC: Fails Management Institute.

Feigenbaum, A. V. (1991). *Total Quality Control.* New York: McGraw-Hill.

Hancher, D. (1991). *In Search of Partnering Excellence.* The Construction Industry Institute. The University of Texas–Austin.

Harrington, H. J. (1991). *Business Process Improvement: The Breakthrough Strategy for Total Quality, Productivity, and Competitiveness.* New York: McGraw-Hill.

Heragu, S. (1997). *Facilities Design.* Boston: PWS Publishing.

Hinze, J. W. (1997). *Construction Safety.* New Jersey: Prentice Hall.

Howell, G., and Ballard, G. (1995). *Improving Performance in Electrical Construction: An Implementation Strategy and Planning System Audit.* Prepared for the National Electrical Contractors Association.

Imai, M. (1986). *Kaizen: The Key to Japan's Competitive Success.* New York: Random House.

Ishikawa, K. (1976). *Guide to Quality Control.* Hong Kong: Asian Productivity Organization.

Juran, J. M. (1988). *Juran on Planning for Quality.* New York: Free Press.

———(1989). *Juran on Leadership for Quality.* New York: The Free Press.

Kerr, W. (1950). "Accident Proneness of Factory Departments." *Journal of Applied Psychology* 34, 167–170.

———(1957). "Complementary Theories of Safety Psychology." *Journal of Social Psychology,* 43, 3–9.

Lemna, G. J., Borcherding, J., and Tucker, R. (1986). "Productive Foremen in Industrial Construction." *Journal of Construction Engineering and Management,* ASCE 112 (2), 192–210.

Leonard, C. A. (1987). *The Effect of Change Orders on Productivity.* The Revay Report. Montreal, Canada: Revay and Associates 6 (2), 1–3.

Liteman, M., Campbell, J., and Liteman, J. (2006). *Retreats That Work: Everything You Need to Know about Planning and Leading Great Offsites.* San Francisco: Pfeiffer.

Loosemore, M., and Teo, M. (2000). "Crisis Preparedness of Construction Companies." *Journal of Management in Engineering,* ASCE 16 (5), 60–65.

MacIntyre, M. (1995). "Facilitator Checklist." *Constructor,* Associated General Contractors of America, Washington D.C., May.

Maloney, W. F., and McFillen, J. M. (1987). "Influence of Foreman on Performance." *Journal of Construction Engineering and Management,* ASCE 113 (3), 399–415.

McGregor, D. (1987). "An Uneasy Look at Performance Appraisals." *Training and Development Journal,* 41, 66–69.

Moen, R., Nolan, T. W., and Provost, L. P. (1991). *Improving Quality through Planned Experimentation.* New York: McGraw Hill.

Moravec, M. (1983). "The Performance Evaluation." *Civil Engineering,* ASCE 53 (11), 55–56.

Moss, S. M. (1989). "Appraise Your Performance Appraisal Process." *Quality Progress,* 22 (11), 58–60.

Niven, D. (1993). "When Times Get Tough, What Happens to TQM?" *Harvard Business Review,* May–June, 20–34.

Oberg, W. (1972). "Make Performance Appraisal Relevant." *Harvard Business Review,* 50 (1), 61.

Olson, R. C. (1992). "The Importance of Planning." *AGC Productivity Improvement Bulletin III,* (17) Associated General Contractors of America, Washington, D.C.

Parker, H. W., and Oglesby, C. H. (1972). *Methods Improvement for Construction Managers.* New York: McGraw-Hill.

Peach, R. W., ed. (1992). *The ISO 9000 Handbook.* Fairfax, VA: CEEM Information Services.

Peters, T. (1987). *Thriving on Chaos: Handbook for a Management Revolution.* New York: HarperCollins.

Riley, D. R. (1994). "Modeling the Space Behavior of Construction Activities." PhD diss. Pennsylvania State University, University Park.

Riley, D. R., and Sanvido, V. S. (1997). "Patterns of Construction Space Use in Multistory Buildings." *Journal of Construction Engineering and Management,* ASCE 121 (4), 464–473.

Robinson, D. G., and Robinson, J. C. (1996). *Performance Consulting: Moving Beyond Training*. San Francisco: Berrett-Koehler Publishers.

Rojas, E., ed. (2008). *Construction Productivity: A Practical Guide for Building and Electrical Contractors*. Ft. Lauderdale, FL: J. Ross Publishing.

Schaffer, R. H., and Thomson, H. A. (1992). "Successful Change Programs Begin with Results." *Harvard Business Review*, January–February, 80–89.

Scherkenbach, W. W. (1991). *Deming's Road to Continual Improvement*. Knoxville, TN: SPC Press

Shetty, Y. K., and Buehler, V. M., eds. (1985). *Productivity and Quality through People: Practices of Well-Managed Companies*. Westport, CT: Quorum Books.

Shewhart, W. A. (1986). *Statistical Method from the Viewpoint of Quality Control*. New York: Dover Publications.

Streibel, B. J., Joiner, B. L, and Scholtes, P. R. (2003). *The Team Handbook*. 3rd ed. Joiner/Oriel.

Thomas, H. R., Jones, J. R., Hester, W. T., and Logan, P. A. (1985). *Comparative Analysis of Time and Schedule Performance on Highway Construction Projects Involving Contract Claims*. Final Report to Federal Highway Administration, Nittany Engineers and Management Consultants. State College, PA.

Thomas, H. R., Sanvido, V., and Sanders, S. R. (1989). "Impact of Material Management on Productivity-A Case Study." *Journal of Construction Engineering and Management*, ASCE 115 (3), 370–384.

Tompkins, J., White, J., Bozer, Y., Frazelle, E., Tanchoco, J. and Trevino, J. (1996) "Facilities Planning." 2nd ed. Wiley, New York.

Vernon, H. M. (1918). "An Investigation of the Factors Concerned in the Causation of Industrial Accidents." Health of Munition Workers Committee. Memo no. 21 (cd 9,046).

Wadsworth, H. M., Jr., Stephens, K. S., and Godfrey, A. B. (1986). *Modern Methods for Quality Control and Improvement*. New York: John Wiley & Sons.

Walton, M. (1986). *The Deming Management Method*. New York: The Berkeley Publishing Group.

Wheelwright, S. and Clark, K. (1993). Managing New Product and Process Development. New York: The Free Press.

INDEX